All Color World of
PREHISTORIC
ANIMALS

All Color World of
PREHISTORIC ANIMALS

Pamela Bristow

OCTOPUS

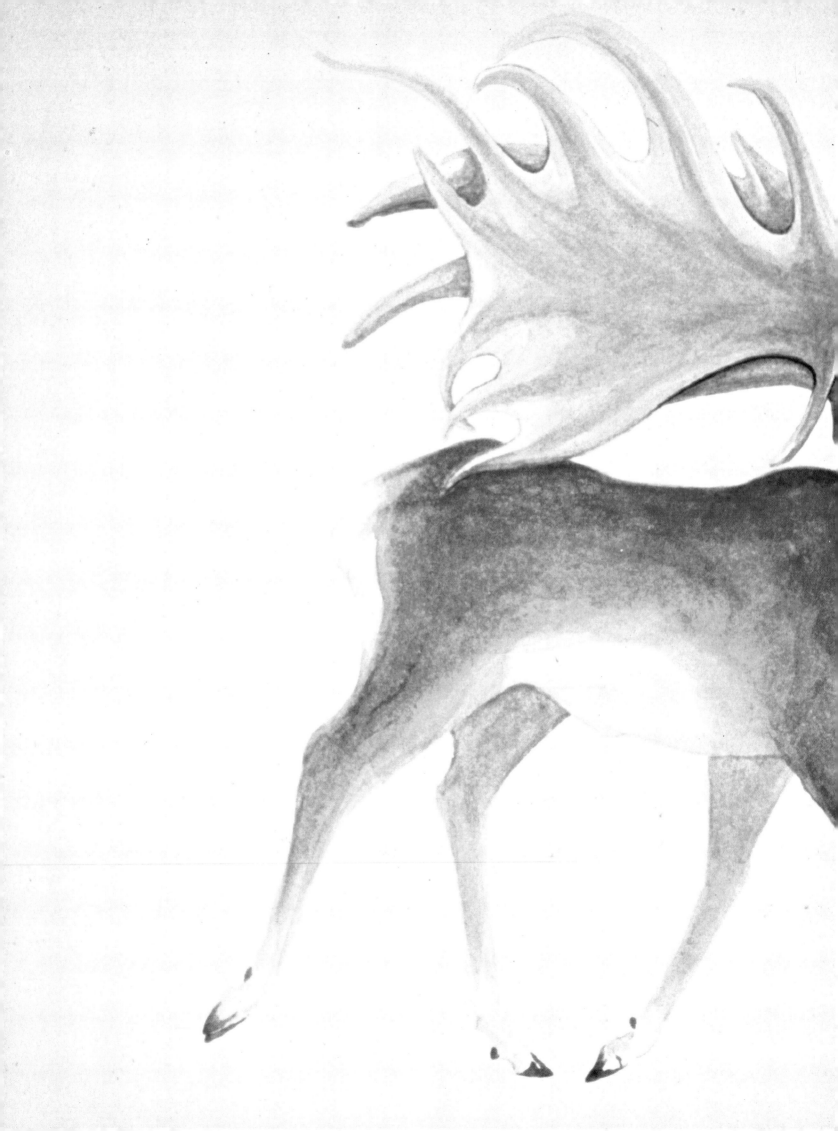

Contents

First published 1980 by
Octopus Books Limited
59 Grosvenor Street
London W1

© 1980 Octopus Books Limited

ISBN 0 7064 1006 8

Produced by Mandarin Publishers Limited
22a Westlands Road
Quarry Bay, Hong Kong

Printed in Hong Kong

Introduction

This planet, Earth, is a fascinating world and its apparently endless variety of living organisms is, to a great extent, responsible for the creation of that interest. Beechwoods, coral reefs, mountain meadows, or ocean depths all have their complement of unique animals and plants; each provides a source of never failing wonder for photographers and small children alike. Living in such diverse environments are some 1.5 million known species of animals, with probably another 3 million yet to be described. Large though these numbers may seem, they are only the present-day representatives of billions of years of evolution, during which time, no doubt, countless organisms lived and died, many without leaving any trace of their existence. Fortunately for us, many developed hard skeletons for protection against enemies or as a support for the softer parts of their bodies, and in rocks from all over the world these hard parts have survived to the present day as fossils. And yet these are mere fragments, often serving only to whet our appetite for the knowledge that lies just beyond our grasp, since these skeletal remains give us few clues as to the lives of the owners and we can only speculate as to the reasons for their disappearance from the Earth.

The greatest mysteries of all, destined never to be totally revealed, are those concerning the origins of our solar system and the beginnings of life on this, our planet, Earth. Scientists today think that, under the influence of gravitational forces, the sun and its worlds condensed out of a cloud of swirling gas. The centre became dense enough to form a shining sun, while local eddies of the gas contracted to form planets. This happened about 4,600 million years ago and it was another 1,100 million years before life began. During this period of time the Earth cooled, the heavier elements, such as iron and nickel, sank to the core of the planet while the lighter elements, such as silicon and aluminium remained in the outer layers. There were large amounts of hydrogen in the original atmosphere, this element being combined with oxygen to form water vapour (H_2O), with carbon to form methane (CH_4), and with nitrogen to form ammonia (NH_3). It was a very different atmosphere to our present one. As the planet cooled, the rocks solidified and the water vapour condensed into rain, which fell in torrents on to the Earth below; the water collected into the basins and formed seas.

The big question is how did life begin in these seas?

Left: Thermal pools such as these in New Zealand may be similar to the rock pools in which life originated.

Right: Early 19th century reconstructions, such as this *Megatherium* by Georges Cuvier, were often inaccurate but signalled a growing interest in fossils.

Overleaf: A painting of *The Deluge* by Francis Danby, 1840. People of the middle ages believed that fossils were the bones of drowned animals.

Modern experimentation has shown that given methane, ammonia and sea water, simple organic compounds such as sugars and amino acids (the units of which proteins are made) can be synthesized under intense irradiation with ultraviolet light – exactly the conditions prevailing in the primordial seas. If the resulting dilute solutions are slowly evaporated, such as would occur in rock pools on the shore, then larger carbohydrate and protein molecules are formed. Gradually aggregations of complex molecules could have produced special droplets (again modern experimentation has confirmed this as possible) which in some ways would have acted like primitive cells. Evolutionary forces acting on these droplets would favour the more stable ones and also those that could acquire the ability to reproduce themselves. Life began. Any date allocated to the origin of life must be to some extent approximate since the whole process took millions of years; however we usually quote 3,200 million years ago as being an average figure, a date far back in the Precambrian era.

This Precambrian era is one of the somewhat artificial divisions of the geologic time scale, a system used to divide the total time span of the earth since it was formed about 4,600 million years ago.

By this system, four eras are recognized; the Precambrian (the earliest), the Palaeozoic (ancient life), the Mesozoic (middle life) and the Cenozoic (latest life). Each era is divided into periods and, in the case of the Tertiary and Quaternary periods, these are subdivided into epochs. The subdivision of the rocks into periods is based primarily upon differences in the contained fossils but it is important to recognize that the scheme is relative. When we speak of the Cretaceous period, for instance, we are referring to rocks and contained fossils overlying those of the Jurassic period, and these Cretaceous rocks are, in turn, overlain by those of the Tertiary period. Of course, Cretaceous rocks were laid down over an interval of absolute time but it is only within the last fifty years or so that absolute dating of rocks has been possible with any degree of precision. This has been made possible by the technique of radioactive dating which depends on the fact that certain elements, such as potassium, decay at a constant rate into another element, in this case argon. Therefore if we know the rate of decay and the proportions of the elements in the rock we are able to estimate an absolute date for the deposition of the rock. In this way it can be determined that rocks laid down at the beginning of the Cretaceous were deposited 136 million years ago, and those at the end of the Cretaceous period 65 million years ago.

The Precambrian era represented the long age, over 4,000 million years, between the origin of the Earth and the appearance of the multicellular organisms. The next great era, the Palaeozoic, lasted for 345 million years and witnessed the origin of all but three of the major plant and animal lineages. It includes the Devonian period, also known as the 'Age of Ferns' or the 'Age of Fishes', and also the Carboniferous or 'Coal Age' when great swamps covered the coastlines. The Mesozoic era, commonly known as the 'Age of Reptiles', lasted for 160 million years and included the time when the great dinosaurs ruled the Earth. During this era the birds and mammals began their evolutionary history, as did the flowering plants. The 'Age of Mammals' is the name given to the last era, the Cenozoic, these animals having undergone great diversification during the last 65 million years.

In Precambrian times, many of the rocks were igneous in origin, that is, they were formed by the cooling of molten lavas. Later, when the weather patterns of rainfall and rivers were established, erosion and sedimentation led to the formation of secondary or sedimentary rocks. These were formed from mud and sand washed into seas and lakes. As debris accumulated, the original deposits were crushed and hardened into rocks by the weight of the later overlying ones. From Cambrian times onward, fossils were quite common in these sedimentary rocks; they were formed when an organism was deposited with the silt at the bottom of the sea or lake, either because it was washed into the water after death or because it lived there. The former process was quite rare and so the fossil record of terrestrial forms is necessarily much less complete than that of the aquatic organisms.

Once the animal or plant was deposited at the bottom of the water, its fate depended on several factors. Rapid bacterial decay would make fossilization highly unlikely, especially if the organisms had no skeleton. However, when a hard skeleton was present, then there were several ways in which fossilization could occur. The remains were often compressed gradually as the rocks formed and hardened and mineralization of the skeleton frequently followed; this process changed the intrinsic bone or shell structure to a more stable mineral form. Part of many rock-forming processes consisted of the chemical formation of 'cement' in the pores between particles of sand or mud; if a potential small fossil, such as a fish or ammonoid, were present, it tended to speed up the cementing process in its immediate vicinity. A structure called a concretion resulted and the fossil inside was often of a high quality and three-dimensional, in contrast to the flattened fossils formed by compression.

Over millions of years seas and lands change. For example, shallow seas and lakes become silted up and vegetation completes the transformation from water to land. Rocks that are formed in deep oceanic basins eventually become part of the land during periods of uplift, a complex process often associated with continental drift. This movement of major land masses is now an accepted theory and explains why the shapes of the continents match so closely. The Rocky Mountains

Above: Undercutting the slab containing a large fossil is an early stage in the excavation process. Parts of this ichthyosaur have already been wrapped and boxed ready for transportation to the laboratory.
Below: The excavated slab has now been splinted and bandaged to reduce the risk of breakage in transit.

and the Andes have been produced by the westward movement of the Americas, while the Himalayas are the result of land buckling, when the Indian and Asian continents collided. When the land is uplifted in this way, then rocks that were previously submerged become part of the terrestrial environment and are therefore subject to weathering and erosion. Through these processes, any fossils are gradually exposed.

Depending on the type of rock and organisms fossilized, fossil collection can be the casual occupation of a summer afternoon or the intensified work of one whole year. Small fossils can readily be collected, often already weathered out of the rocks, or easily extracted with a geological hammer. Concretions are frequently found in this way and can be split open to reveal the contained fossil, which lies along a plane of weakness. If the concretion is calcareous and the fossil siliceous (a common combination), then the three dimensional aspect of the skeleton can be revealed by dissolving the rock in acetic acid.

Larger, compressed specimens, such as articulated skeletons of dinosaurs (those with the bones still connected), demand a high degree of skill if they are to be removed from the rocks without damage. Often only the tip of a skull or limb-bone provides a clue as to the position of the animal and so the first task is to expose the bones, which are then covered with plaster-soaked layers of cloth; the articulated skeleton can then be excavated intact, since the plaster provides the necessary support. The plastering process is then completed on the other side of the specimen to provide protection for the journey to the laboratory. Once there, the plaster is carefully chipped away, together with any remaining rock, and the specimen is mounted ready for study and display. Clearly, this kind of job is one for a specialist.

Of course, fossil collection and preparation are simply mechanical tasks. They can be exciting in themselves, for there is always the chance of a major find just around the next outcrop of rock. Preparation of specimens can be tedious but there is the satisfaction of a job well done at the end and there is always the possibility of some new feature of interest appearing in the process. But the great excitement of palaeontology lies in the knowledge to be gained of bygone ages. For the past of the Earth is our past, the organisms lying in the rocks are part of our history and the worlds revealed by those ancient remains are strange and fascinating indeed.

Era	Period	Epoch	Beginning (in millions of years before our our age).	A selection of plants and animals that were significa (This chart shows isolated examples only, and doe:
CENOZOIC		HOLOCENE	0.01	Orchid · Blue Whale · Modern Man
	QUATERNARY	PLEISTOCENE	2	Smilodon · Australopithecus
	TERTIARY	PLIOCENE	7	Alticamelus · Pliomastodo
		MIOCENE	25	Grass · Desmostylus · Dinotheriun
		OLIGOCENE	38	Astrapotherium · Arsinotheri
		EOCENE	54	Oak Tree · Basilosaurus · Uintatheriun
		PALAEOCENE	65	Paramys · Eurymylus
MESOZOIC	CRETACEOUS		136	Conifer · Corythosaurus · Allosaurus
	JURASSIC		190	Cycas · Ichthyosaurus · Rhamphorhync
	TRIASSIC		225	Cynognathus · Coelophysis
PALAEOZOIC	PERMIAN		280	Eryops · Dimetrodon
	CARBONIFEROUS		345	Lepidodendron · Meganuera · Eogyrinu
	DEVONIAN		395	Fern · Osteolepis
	SILURIAN		440	Cooksonia · Climatius · Crinoid
	ORDOVICIAN		500	Nautiloid · Coral
	CAMBRIAN		570	Alga · Trilobite · Lingula
	PRE-CAMBRIAN		4,600	PROTEROZOIC Formation of Earth 4,000 milli

Floras and faunas			Significant life-forms	Significant climatic and geologic events
Seagull — Kangaroo		**AGE OF MAMMALS**		
Glyptodon — Megatherium			Rise of Man.	Northern ice cap spreads far to the south– THE ICE AGE.
Synthetoceras			First early Man-like primates.	Polar glaciation began in Northern Hemisphere. Present day ocean currents become established.
Proconsul — Merycodus			Spread of grasslands causing change in mammal fauna	Mountain building resulting in the Alps. Continents close to their present position.
Cynodictis — Bronto-therium			Large mammals common.	Cooling of climate with consequent regional extinction of tropical forests.
Diatryma — Eohippus			Diversification of many mammal groups. Flowering plants become common.	Climate generally equable. Tropical and subtropical conditions existing in what are now temperate areas.
Oxaena — Phenacodus				
Triceratops — Pteranodon		**AGE OF DINOSAURS**	Extinction of dinosaurs and some other reptile groups. Ammonites die out.	Clear seas deposit chalk in many parts of the world. Climate becoming more seasonal.
Brontosaurus — Archaeop-teryx			'Age of Cycads'. Hey day of the giant dinosaurs.	Super land mass–Pangaea begins to break up into separate continents. Shallow seas encroach upon the land masses.
Rutiodon — Turtle			Origin of mammals, dinosaurs and modern-day reptile groups.	Deserts common; climate generally warm.
Moschops — Diplocaulus			Many invertebrate groups become extinct in seas. First mammal-like reptiles.	Continents forming one super landmass –Pangaea. Climate equable and warm; many deserts in hinterland.
Lamellibranch			Extensive coal swamps– 'The Coal Age'. First reptiles.	Europe and North America straddled the Equator and were covered with lush vegetation. Southern Hemisphere. continents largely covered by ice.
Coccosteus — Cheirolepis			Plants establish themselves. Strange fishes inhabit seas, rivers and lakes.	Soils forming following invasion of land by plants. Large inland lakes and extensive estuaries and deltas.
Graptolite			Plants begin to invade the land. Graptolites common in the sea.	Continental areas largely barren, with little soil. Most life still confined to the sea.
Brachiopod — Trilobite			Origin of very primitive fishes.	Following a late Precambrian Ice Age, temperatures and sea level rose. Low lying coastal areas become covered by sea. Towards the end of the Ordovician a huge ice cap returned to cover the South Pole, then situated in the present day Sahara Desert.
Crinoid — Gastropod			Appearance of animals with hard skeletons.	

ars ago

Early Seas

For many years, the Cambrian period was considered to be the one in which life originated on this planet. We know now that this was not so, that life had been evolving for at least 2,900 million years before the onset of the Palaeozoic era. The incredibly ancient, Precambrian life forms were generally unicellular and, lacking hard skeletal parts, they were rarely fossilized. In late Precambrian rocks a few worms and jellyfish have been found; and then, over a time span of only 20 million years (a fraction of a second on the geological clock), a great multitude of animals appeared. These organisms were mostly multicellular and had chitinous or calcareous skeletons, which made them ideal subjects for fossilization.

What were the differences between Precambrian and Cambrian world climates and seas, that made this explosion of life possible? Experts have suggested that gaseous oxygen may have reached a level where an ozone layer formed in the upper atmosphere, providing effective protection to the organisms on the planet surface against harmful exposure to ultraviolet light. Higher dissolved oxygen levels in the seas would have

Left: Pterygotus **was a particularly large and ferocious eurypterid, being between 2–2.5m (6.5–8ft) in length.**
Below: **The coral,** *Lonsdaleia,* **was an important contributor to the reefs of the Carboniferous times.**

provided a tremendous boost to the activity of the plants and animals. Other people think that the increase in animal populations can be explained by the advent of the skeleton. This may have been due to a change in marine chemistry making calcium carbonate much more freely available. Of course, at this time in the Earth's history the land was barren, devoid of life; what organisms there were lived and died beneath the waves. Gradually, during the Cambrian period, extensive areas of land became inundated by shallow sunlit seas, thus greatly increasing the areas where animals and plants could live. Their rapid multiplication could well have been a response to vast new seas in which they could thrive.

Amongst the animals that quickly became abundant were the brachiopods and trilobites. Both lived at the bottom of the shallow seas, the brachiopods apparently limited down to a depth of 200m (650ft), while some of the trilobites also inhabited deeper waters down to about 600m (1,900ft). Initially they both had chitinous exoskeletons but anatomically they had little else in common.

The primitive brachiopods were enclosed in two unhinged chitinous plates and were sessile, attached to the sea-bottom by a pedicel or stalk. Later Cambrian forms developed calcareous hinged shells and were

Left: The trilobites were common throughout the Palozoic era, becoming extinct at the end of the Permian period. They were called trilobites because their bodies were divided longitudinally into three sections, a central column and two lateral lobes. This specimen, *Griffithides*, has a highly ornamented head shield and relatively small eyes.

Right: Crinoids have been found in clear seas since Ordovician times. This specimen has a stalk and the five-rayed symmetry of the group to which it belongs, the echinoderms. It is seen here with two reef building corals.

free-living or cemented to the sea-floor; these two forms of skeleton were indicative of a basic diversification of the brachiopods and the two resultant groups are known as inarticulate (unhinged) and articulate (hinged). The animals were filter-feeders, using a complex ciliated organ called a lophophore to create a current that carried oxygen to the respiratory surfaces and unicellular algae to the mouth. Like modern bivalves (to which they are not related), the brachiopods were gregarious, living in such crowded conditions that the shells were often damaged; fossil evidence of abrasions and subsequent repairs is quite common.

The trilobites, in contrast, lived solitary lives wandering over the sea-bottom and they retained a jointed, chitinous exoskeleton throughout their evolutionary history. They were segmented arthropods with jointed appendages on each segment and they used these appendages both for walking and digging and probably also in feeding. Some of them created water currents and were filter feeders while others captured small prey or scavenged in the detritus on the sea-floor. Their method of breathing is subject to a certain amount of controversy, some authors suggesting that the appendages were utilized for this function, while others believe that respiratory exchange took place through the thin ventral surface.

Like modern arthropods, trilobites grew in a series of jumps, at each stage going through a process of moulting; some of the fossils that have been found represent shed exoskeletons, rather than whole animals. The largest specimens were about 67cm (27in) long and these relatively huge forms were probably immune to most predators. However, many trilobites had defence mechanisms which can only have been aimed at potential enemies. They had large eyes which, like insect eyes had several thousand separate lenses and presumably a mosaic image was formed. Using these sense organs to detect enemies, a trilobite could then employ its second line of defence; it rolled into a ball like a woodlouse. Fossils are often found in this position.

The eurypterids could well have been the fiercest enemies of the trilobites. Early forms were quite small, but by the Devonian, some of them had reached 200cm (6ft) in length – the largest arthropods ever known. They looked like giant aquatic scorpions and they were probably distantly related to those poisonous animals. Many eurypterids carried a pair of large crab-like claws, and these have caused experts to believe that they were highly predatory forms; the claws may also have been used in defence against other eurypterids and against predatory fishes in later Palaeozoic times. Eurypterids were amongst the first animals to leave the sea and they have been found in many estuarine and freshwater deposits, both in North America and Europe; the marine forms are often found closely associated with trilobites and brachiopods.

The two latter groups continued to increase in both numbers and species throughout the Cambrian and they reached a peak during the Ordovician marine encroachment on North America. Thereafter their numbers declined until widespread extinction overtook them and the eurypterids in the Permian. The trilobites and eurypterids died out completely, but the brachiopods survived and went on to invade the sea floors where they have lived to the present day.

During the Ordovician period, a phenomenon began that changed the lives of many sea dwellers. Throughout Cambrian times the corals had been solitary or primitively colonial, but in the Ordovician

some of them became major reef builders. These reefs were situated in the littoral zones and originated when an individual reproduced asexually and the resulting offspring did not separate from the parent. Only the outermost layer of a reef contained living polyps, each enclosed in a calcareous cup, and the inner and lower regions were built up from the cups of dead individuals. Each outermost layer gave rise to the next before it died, and in this way the reef could become, in time, a massive structure. These corals were fairly demanding in their environmental requirements, living only in clear, shallow water to a maximum depth of 50m (160ft); in this zone the sea was well illuminated by sunlight and wave action was a constant factor, ensuring a steady supply of well-oxygenated water around the reef.

The Palaeozoic corals were different in structure, although similar in life styles, to modern forms and at the end of the Permian, the majority of them became extinct. Only a few survived to give rise to an entirely new coral group, from which the forms which exist today have evolved.

Brachiopods were common inhabitants of the early seas. They had two hinged valves and in this specimen these were ornamented both by radiating lines and by concentric folds.

18

Reconstruction of a shallow Ordovician sea. In the foreground are a straight-shelled nautiloid and a trilobite, while in the middle ground are groups of tube-dwelling worms and corals together with a curved-shell nautiloid. Attached to the sea-bed in the background are crinoids and algae with jellyfish swimming above them.

Living on the muddy bottoms of streams and lakes were fishes such as *Cephalaspis salweyi*. This particular ostracoderm comes from the Old Red Sandstone of Western England, from rocks of Devonian age.

Of course, the reefs provided shelter for many animals and plants, amongst which were worms of all kinds, sponges and snails. One common group of animals, living for the most part on the sheltered side of the reef or on the sea-floor, were the crinoids or sea-lilies. These organisms originated in the Ordovician and were members of the group known as echinoderms or spiny-skinned animals. Most echinoderms were freeliving, such as the starfish and sea urchins but a sea-lily was attached to a substrate, such as a reef or boulder, by a stalk on the top of which was a calyx or cup enclosing the body. Both the stalk and the calyx were covered by calcareous plates and the calyx had a ring of branched arms on its outermost edge; these arms were responsible for the creation of feeding and respiratory currents.

A fascinating association has been documented between certain Palaeozoic crinoids and a group of snails, which lived between the arms of the crinoid calyces and ate leftover food. The snail shells became greatly modified during the course of this relationship, the early ones having typical coiled shells, while those of later forms became uncoiled, the better to fit between the arms. These particular crinoids became extinct in the Permian and the snails also failed to survive, thus demonstrating their total dependence on their hosts.

Ever since snails originated in the early Cambrian time, they have been a highly successful group; originating in the sea, they have invaded freshwater and terrestrial environments, grazing their way across the algae and other vegetation, wherever soft young growth can be found. Representatives of the earliest snail groups are still living today and, although primitive in some respects, they appear outwardly like many other snails. The bivalves, for example oysters, mussels, clams and the like, were another group of molluscs that originated in the Cambrian, and again the early forms looked much like modern representatives. Both the snails and the bivalves lived at the bottom of the sea, the snails crawling slowly around, while the bivalves were either anchored to the substrate or buried in the mud and sand. Some bivalves, like modern oysters, were found crowded together in great numbers and their remains have formed huge shell banks in the rocks.

The other great group of molluscs found in the Palaeozoic were the cephalopods, represented then by the nautiloids and ammonoids, and today by the octopus, squid and nautilus. The former groups lived a very different kind of life to that of the snails or bivalves; they had coiled, chambered shells in which some of the chambers acted as buoyancy tanks. They swam or floated in the warm, upper waters of the seas, feeding on drifting crustaceans and other small organisms. The nautiloids originated in the Upper Cambrian and were probably ancestral to the ammonoids which appeared during early Devonian times.

20

Drepanaspis was a primitive jawless fish from Devonian times. It was an ostracoderm, about 30cm (12in) long and had the typically flattened form of the bottom dweller where it lived by sucking up mud.

This latter group was just beginning its evolutionary history in the Palaeozoic and its heyday was to come in Mesozoic times; however, along with so many other animals, the ammonoids became extinct at the end of the Cretaceous. The nautiloids were never as abundant as their relatives, but the nautilus has survived to the present day.

The ancient seas also provided an environment for early fishes, but unlike the nautiloids, these vertebrate animals (those with backbones) preferred shallow, coastal waters, where they spent most of their lives resting on the sea-floor. These primitive fishes appeared in Ordovician times and were quite different from the later fishes in that they had no paired fins (the fins which, in most fishes, correspond to our arms and legs), nor jaws. Instead, the mouth was a simple circular opening on the underside of the head. Mud-grubbing was their trade; they either scooped or sucked up mud and allowed the gut to extract the organic matter. Swimming must have been a rather inefficient affair, for not only did they lack the paired fins, but they were also heavily armoured with a large, bony shield over the head and bony scales covering the trunk and tail. This has given rise to their name, ostracoderms, which means 'shell-skinned'.

Most of these forms were 'tadpole-shaped' and they rarely exceeded 20cm (8in) in length. However, one group, the cephalaspids (meaning 'bony-headed') were shaped rather differently with flattened, semi-circular head shields. They inhabited brackish waters of the late Silurian and early Devonian. One curious feature – the presence of special, scaly areas, the lateral fields, on the otherwise solid head shield – has attracted a degree of speculation, for there are no modern fishes with anything comparable. Some people think that they were electric organs used either for defence or as a navigational aid in the murky estuarine waters, while others believe that they were devices for detecting underwater sound waves. We do not know the answer, but the lateral fields were clearly important in the lives of these quaint-looking fishes.

The ostracoderm reign was a short one, for the end of the Devonian saw their extinction, some 150 million years after their entrance. They have left two jawless descendants, the lamprey and the hagfish, which survive in a world of jawed fishes by adopting very specialized feeding habits; the lamprey is parasitic, while the hagfish literally eats its way through dead or dying fish.

The early seas are fascinating for their very unfamiliarity. Strange-shaped fishes vie with the trilobites or eurypterids for our attention, while even such apparently familiar structures as the coral reefs were really quite different in many ways. Imagine a coral reef without fish and one of the major changes will immediately be apparent. The very remoteness in time of these seas ensures their continuing fascination, for nothing comparable is likely ever to occur again.

Experiments and Transitions

The Palaeozoic era was one of the most important times in the history of life on this planet. The early part of the era was a time of experimentation when the majority of the animal groups appeared and the great invertebrate and fish lines of descent were established in the seas. The later Palaeozoic was a time of great transition when some plants and animals left the water and began the conquest of the land.

We have already seen how the jawless fishes and

Left: Labyrinthodonts were primitive amphibians from Carboniferous and Permian times. They probably lived in swamps where their bodies were supported by water. *Below: Pteraspis* was a Devonian ostracoderm of the type known as a heterostracan, fishes characterized by stout armour covering the anterior part of the body and by a reversed heterocercal tail.

invertebrates evolved and how a multitude of grotesque forms inhabited the early seas. The first of the jawed fishes were also strange when they appeared in the late Silurian, and the time of experimentation continued throughout the Devonian. In fact, the Devonian has often been called the 'Age of Fishes', for it was this interval of time that witnessed a considerable increase in the fish numbers and a great diversification of their ranks. The placoderm or plate-skinned fishes would have appeared bizarre to our eyes, for there is nothing comparable living today. They were almost exclusively Devonian, inhabiting coastal waters, rivers and lakes where they spent their time on or near the bottom. The head and trunk were each encased in a bony shield, while between these two

shields there was a neck joint, which allowed the head to be rotated upwards during feeding, thus increasing the size of the mouth opening. Some, such as *Coccosteus*, were clearly predaceous types in which the jaws were armed with great shearing blades for severing prey; others, such as *Dunkelosteus*, were veritable giants, growing to a length of 7m (23ft). However, the majority rarely exceeded 65cm (25in). Yet other placoderms were superficially similar to modern rays and presumably remained buried in soft sand, stirring only to feed on hard-shelled invertebrates. By the end of the Devonian, the placoderm reign was all but over and we can only assume that competition from more efficient forms such as sharks and bony fishes became too severe.

The latter were becoming very numerous as the Devonian period wore on. In contrast to the placoderms, early bony fishes were inhabitants of surface waters. They had large eyes suggesting that sight was an important sense in their lives, while placoderms were much more dependent on their sense of smell, judging by the size of their nostrils. This difference was probably associated with their differing modes of life. Early bone fishes were encased in a thick, bony armour but, instead of two rigid shields as was characteristic of the placoderms, they had a series of small plates over the head and a great many rhomboid scales over the body. So, combined with protection, there was also considerable flexibility between both the plates and the scales – a sort of 'chain-mail', as against a 'suit of armour'. It is also assumed that these early bony fishes had a gas-filled swimbladder (as do most modern bony fish) which acted as a buoyancy tank and this innovation allowed a considerable saving of energy, obviating the need for constant swimming to maintain their level in the water. These early, bony fishes were called palaeoniscids, because of their ancient scale structure, and were in themselves rather insignificant (most were barely larger than minnows), but they were important as the forerunners of an enormously successful lineage, which has resulted in the astonishing diversity of bony fish living today. They number about 19,000 species, equal to the rest of the living vertebrates put together.

While these strange, early fishes swam the Devonian seas, other life-forms were beginning the greatest conquest of all. For eons of years the waters had sustained life, but the land had been barren. Literally and completely barren, it had consisted of rocks and mountains, icecaps and glaciers, but there was nothing that could be described as soil; and without soil no life could become established. However, by the Devonian,

Left: This reconstruction of a Devonian landscape shows early clubmosses (*species of Protolepidodendron*), psilophytes (*Psilophyton*), and ferns (*Cladoxylon*) in the foreground. Other species of clubmosses in the background are more tree-like.

Below: One of the best known Devonian fishes was *Eusthenopteron* which inhabited freshwaters approximately 350 million years ago. It was about 45cm (18in) long and must have led a life-style similar to that of a modern pike.

the first tentative steps had been taken towards the invasion of that whole new terrain. Early land plants were present in lower Devonian times, but whether they grew in a swamp or whether terrestrial soil formation had been initiated by bacteria and fungi, we do not know. Inevitably though, the slow invasion of the land by the plants themselves ensured the formation of soil.

Zosterophyllum, one of the first land plants, is found in lower Devonian rocks of Scotland. It consisted of a tangle of 'rhizomes', from which grew a tuft of dichotomously branched stems. This form of branching is seldom seen in modern plants, but was very common in primitive ones; each stem branched into two at one point and each of the sidebranches in turn branched into two and so on to form a kind of candelabra. Sporangia (spore-bearing organs) were found on these aerial stems and the plant reproduced itself by this means. *Rhynia* was a similar plant from the middle Devonian of Scotland and its remains, together with those of related species, form a petrified peat bog fully 2.5m (8ft) thick. Obviously, by middle Devonian times the conquest of the land was well on the way.

By the end of the Devonian, early representatives of the clubmosses, calamites and ferns were established in many areas of the world. Australian, European, Russian and North American rocks have all yielded fossil plants of this age, and from the northern hemisphere localities have come, in addition, *Callixylon*. This plant provides a classic example of the difficulties facing palaeobotanists. The name had been given to a plant with the dimensions of a tree, being up to 20m (60ft) high. Only the trunks had been found and, since they resembled those of modern conifers, it seemed a reasonable supposition that this was an early form. Other plant material was found at the same localities, but no conifer-like seeds were present and no leaves had been found attached to the trunk. Plants usually disintegrate into stems, roots, leaves and seeds (or spores) after they die and these components are therefore fossilized separately. Until a connection can be demonstrated between the different parts, no-one can be sure which leaf fits which stem; but we can make intelligent guesses and it was thought that the leaves of *Callixylon* would be needle-shaped, like those of a pine. Imagine the surprise of the experts then, when a trunk was found with leaves which proved to be large fern-like fronds attached. So was

From the Old Red Sandstone of Scotland come these two examples of early jawed fishes. The hunter, *Coccosteus*, was about 65cm (25in) long and belonged to the placoderm fishes known as arthrodires, a group which dominated the streams and lakes of late Devonian times. The hunted fish, *Diplacanthus*, was an acanthodian, a group of fishes characterized by a whole series of paired fins running along the length of the body.

Callixylon an early conifer or an enormous fern-like plant? Neither suggestion is completely acceptable, for conifers do not have fronds and ferns do not have conifer-like trunks. Definite evidence as to whether the plant bore conifer-like seeds or fern-like spores would resolve the question.

Large areas of the late Devonian landscape were covered with vegetation. Trees, such as *Callixylon*, grew tall and shaded the ferns and horsetails, which crowded the edges of the pools and slow-moving rivers; the growing plants accumulated mud and silt at their bases and so marshes and swamps formed and spread. The scene was set for the emergence of the backboned animals on to land.

This major transition must have occurred during the upper Devonian. At that time there were two groups of similar animals; on the one hand were the rhipidistian (tassel-finned) fishes, while on the other hand were primitive amphibians. Anatomically they were closely similar, implying that their common ancestry was not too far removed.

Rhipidistians were true fishes; some, such as *Osteolepis* swam the Devonian lakes, while others, such as *Megalichthys*, inhabited the coal swamps of the Carboniferous period. Fish though they were, their skeletons showed several land dwelling adaptations, these being chiefly designed to overcome the increased effect of gravity on land as opposed to in the water. *Eusthenopteron* was a well-known Devonian rhipidistian, about 45cm (18in) long, with a plump body. The paired fins, corresponding to the amphibian fore and hind limbs, were particularly interesting: outwardly they were fish-like, but the internal skeleton was very strong and there were bones corresponding exactly to the upper and lower segments of the amphibian limb skeleton. Clearly, these limbs would have afforded more support to the body than the typical, flimsy fish fin. It is possible that *Eusthenopteron* undertook short excursions on to land. The vertebral column was also similar to that of a primitive amphibian. Each vertebra was a stout, bony element with incipient processes interlocking with those on adjacent vertebrae. The result was a backbone which could resist sagging when the animal left the water, an important attribute in a land vertebrate.

In the roof of the mouth there were two internal nostrils, as found in amphibians and other tetrapods. This suggests that *Eusthenopteron* had the capacity to breathe air, using lungs, as well as obtaining oxygen through typical fish gills. Geological evidence suggests that this animal may have lived in freshwater environments periodically subject to low oxygen regimes, and air breathing would have been distinctly advantageous. In many ways *Eusthenopteron* was still a fish. Its overall shape was like that of a pike and we may imagine that it led a similar life, lying hidden in water vegetation, ready to rush out and seize unwary prey.

It is probable that the primitive amphibian, *Ichthyostega*, which was about 90cm (35in) long, and many of its relatives, lived much of their lives in water. They had, however, lost all trace of gills as adults, relying exclusively on air breathing. The limbs were stronger and, significantly, the pelvic girdle (or hip girdle) was anchored to the vertebral column, allowing the locomotory thrust of the hind legs to be transferred efficiently to the rest of the body. *Ichthyostega* could raise the body clear of the ground, even though its sprawling gait may seem primitive to our eyes. Indeed, many of the more advanced amphibians retained this rather ungainly locomotion, and most of them must have been more at home in water or in a swamp than on the land.

Ichthyostega, a primitive amphibian grew to about 90cm (35in) long. The wide mouth, the eyes situated near the top of the head and the sprawling gait are characteristics of these early tetrapods. It is probable that *Ichthyostega* spent much of its time in water.

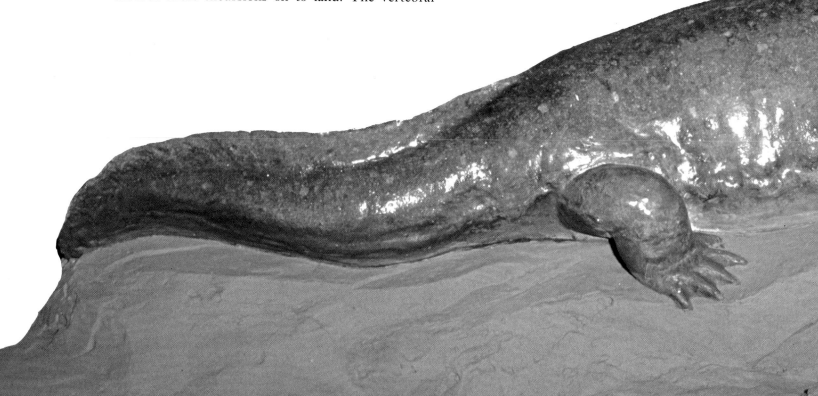

Above: Closely related to the rhipidistians (the group of fish from which amphibians evolved) were the lungfishes or dipnoans. *Dipterus* was a typical Devonian lungfish with heavy scales, two dorsal fins and a heterocercal tail. It is possible that *Dipterus* had functional lungs similar to those found in modern lungfish.

Inefficient though the early amphibians were, their descendants survived into Permian times and some of them even became reasonably well adapted for life on land. *Eryops*, a Permian labyrinthodont, was a heavy animal, about 180cm (6ft) long, that preyed upon fish and smaller amphibians. The skull was flattened, an adaptation seen in animals living in shallow water; however, it had sturdy limbs and feet and a strong vertebral column, and could probably have made good progress on land.

Unlike these 'terrestrial' forms some of the amphibians never left the water, while others returned to it. One interesting 'group', the branchiosaurs, which lived in late Carboniferous and Permian times, were thought to have been an evolutionary side-line that had adopted such an aquatic existence. However, it is now known, from their small size and gill skeletons, that they were larval forms, unrelated to each other and equivalent to our modern tadpoles. A whole series of developmental stages has been found from the smallest 'tadpole' to a nearly mature *Eryops*-like animal. Aquatic amphibians were common in the Permian, for by this time the reptiles were establishing their dominance on the land and the water became the only medium in which the amphibians could thrive. Some of these aquatic forms evolved from more truly terrestrial forebears; others, such as *Diplocaulus*, were the descendants of amphibians that had never left the water. *Diplocaulus* was a highly distinctive form, in which the skull bones grew sideways to such an extent that the head of the animal appeared

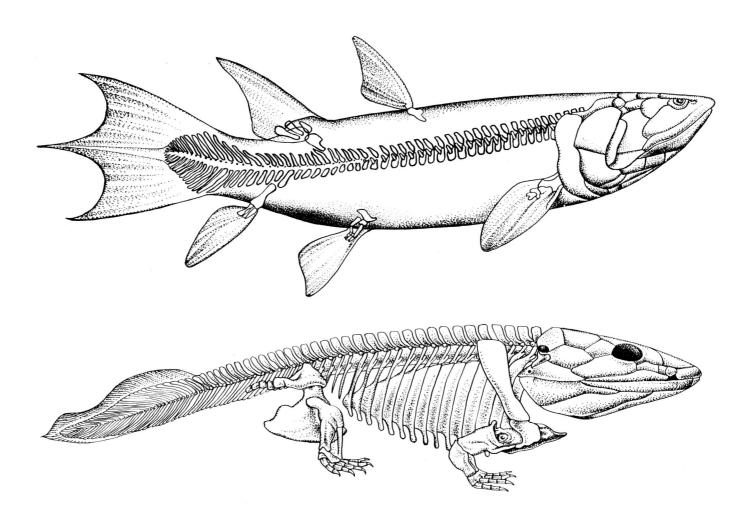

like the point of an arrow. For many years this strange adaptation puzzled many people. However it has recently been suggested that it may have acted as a hydroplane, aiding rapid swimming.

No description of the transition from water to land would be complete without some mention of the Carboniferous swamps, for it was there that much of the transition took place. Resemblances to a modern swamp were few; most of the plants were strange by our standards, but a few have survived in archaic form to give an inkling of the appearance of their ancestors in their heyday. The small horsetails that grow in marshlands and hedgerows are such survivors; with their odd jointed stems and sheath-like leaves, they appear peculiar and primitive, when compared with the flowering plants. The ancient horsetails of the late Carboniferous swamps formed a large part of that flora, growing up to 30m (100ft) high, even though the stems were quite hollow; each stem bore up to 60 whorls of leaves. With the same stiff, jointed appearance as their modern representatives, they must have looked quite unnatural.

The modern clubmosses, like the present-day horsetails, are small, creeping plants, but their late Carboniferous forebears were as striking as the horsetail ancestors, albeit in quite a different way. They were trees, growing as much as 40m (130ft) in height,

A stout vertebral column, rib cage and strong sprawling limbs distinguish early fishes such as *Eusthenopteron* (*above*) from the primitive amphibian (*Ichthyostega*).

with a crown of leafy branches. They had no true roots, but instead had four underground 'rhizomes' that branched dichotomously to form a horizontal network.

Neither the giant horsetails nor the clubmosses were fully adapted to life on land, since they produced spores which needed water to develop and form the next generation and since also the water transport systems in their stems were relatively inefficient. The cordaites, by contrast, produced seeds instead of spores and these could develop in soil with a minimum of water. They were trees, growing up to 40m (130ft) high, and water was transported up the trunk in wood, a technique that is still used by modern trees. Obviously these plants were much better adapted to terrestrial life than their horsetail or clubmoss contemporaries, and it is likely that the cordaites forests could grow on relatively dry ground.

As in any present day swamp, the undergrowth was as important as the trees. A large proportion of the ground cover plants were ferns, many of which could have been mistaken for their modern counterparts; then, as now, they produced spores and so needed water to complete their life cycles and ensure the next

Diploceraspis was a Permian amphibian which, like *Diplocaulus*, had an arrow-shaped skull; this strange adaptation may have aided rapid swimming.

generation. Among the low-growing ferns, another group of plants grew and thrived. Superficially similar, they were actually very different – they had hard, woody stems and they produced seeds, not spores. Often called the seed-ferns, some of them bore crowns of seed-bearing fronds on the tops of short trunks. Others were vine-like in form, probably trailing and climbing, and using the trees and ferns as supports.

As the wet Carboniferous period gave way to the drier Permian time, the primitive spore-bearing plants were gradually replaced by more advanced seed-bearing forms; at the same time, diversification of the reptiles created significant changes in the terrestrial animal populations. The earliest reptiles differed from their amphibian ancestors in many important ways; for instance there were notable differences in almost all parts of the skeleton, mostly concerned with a greater degree of adaptation to terrestrial life. Some of the differences between modern amphibians and reptiles can only be inferred for their extinct representatives, because they do not involve skeletal features. For example, we assume that the skin of both primitive and modern amphibians was moist and lost water very rapidly, in contrast to that of reptiles,

which was covered in horny scales to reduce excess evaporation. We also think that early amphibians like the modern frogs and newts laid many small eggs in water, whereas the reptiles produced relatively few, large eggs, which could develop on land.

From the very early reptiles, several lines of descent soon appeared, amongst which were the mammal-like reptiles, destined to give rise to the mammals. The primitive members of this group, the pelycosaurs, were typical reptiles in most respects, but several species evolved an enormous 'sail' on the back, which has puzzled experts for many years. In *Dimetrodon*, the 'sail' was formed by extensions of the vertebral spines, and we think it very likely that there was a web of skin stretched between the spines. In another animal, *Edaphosaurus*, there was a similar 'sail' but the supporting spines were stouter and linked by interconnecting crossbars. *Edaphosaurus* and *Dimetrodon* were but distantly related, the first belonging to a herbivorous lineage and the latter being a carnivore, therefore their 'sails' were evolved independently. Many theories have been forwarded as to the function of the 'sails', ranging from use in threatening behaviour and defence, to being a male characteristic. The most commonly accepted explanation is that the 'sail' was a thermo-regulatory device. If the owner turned to face the sun, heat would be lost from this

31

large expanse of skin; if the 'sail' was turned sideways on to the sun then heat would be absorbed.

Not all the pelycosaurs had 'sails', and from some of these other animals evolved a second group of mammal-like reptiles, the therapsids. From this lineage were derived the theriodonts, the ancestors of the mammals. *Cynognathus* was a typical, if somewhat large, member of this assemblage; about 2m (6ft) long, lightly built and carnivorous. It had several features which were not found in typical reptiles, but which were characteristic of the mammals: for instance, a secondary palate was developed in the roof of the mouth. This plate of bone separated the mouth from the nasal cavity and enabled the animal to eat and breathe at the same time. The stance was decidedly mammal-like, for the limbs were directly beneath the body with the knees pointing forwards and the elbows backwards. Another mammal-like feature found in this animal was its differentiated dentition; in contrast to the essentially similar teeth of reptiles, it had incisors at the front of the jaws, canines or eye teeth, and a set of cusped, cheek teeth. We do not know, of course, whether the theriodonts had hair or bore their young alive, and since the trends throughout their evolution were towards an ever more mammal-like form, it has proved difficult to know where the reptiles ended and the mammals began.

Above: Cynognathus, a Triassic theriodont, was an active carnivore with sharp eye teeth. Its legs were tucked beneath the body and this adaptation helped it to run down its prey.

Right: One of the strangest pelycosaurs was *Dimetrodon* from the Permian of North America; the enormous sail on its back probably functioned as a temperature regulation device.

This apparently insoluble problem has been given a somewhat arbitrary solution involving the jaw articulation; different bones were involved in this joint in reptiles and in mammals, and the point at which the change took place has been judged to mark the boundary between the two groups of vertebrates.

The transition on to land had by now become a *fait accompli*. The mammal-like reptiles were, in themselves, well adapted to a terrestrial existence and the mammals, when they appeared during the course of the Triassic, completed the transition. At the same time another lineage, descended from the early reptiles, was evolving rapidly until, again in the Triassic, the terrestrial dinosaurs began their diversification which was to continue so successfully throughout the Mesozoic. From this group of reptiles came the birds, the vertebrate group that has so completely conquered the air, an even more difficult medium than the land.

The Age of Dinosaurs

No other interval in the earth's history has such charismatic appeal as the Mesozoic era, that time between 200–60 million years ago, which belonged to the reptiles and in particular to the dinosaurs. To anyone remotely interested in prehistoric life, dinosaurs hold a special fascination; the very word, which literally means 'terrible lizard', conjures up mental images of dragons, so popular in medieval myths and legends. However, dragons they were not; they did not breathe fire but they were certainly awesome in their own way, especially with regard to their size. Some of the herbivores reached over 27m (90ft) in length, and even some of the carnivores exceeded 13m (42ft). Their size must contribute greatly to their popularity, but perhaps their greatest appeal lies in their total extinction – if we had seen them we might have accepted them as commonplace, but an interval of some 50 million years separates the last of the dinosaurs from the first of the men.

The world inhabited by the dinosaurs was rather different to that of those first men. For instance, the climate of the world was rather warm and equable throughout the Mesozoic era, providing ideal conditions for reptile existence. Consequently the dinosaurs spread worldwide, their remains being especially common in North America and Asia. They inhabited the banks of huge rivers, low lying marshland areas and coastal lagoons. During the Jurassic period the lush vegetation of these areas was composed of cycads, ferns and conifers. The first of these plant groups was very common and has caused botanists to name the period the 'Age of Cycads'. During the Cretaceous time a major vegetational change occurred with the advent of the flowering plants, and the cone-bearing cycads and conifers were largely replaced by more familiar palms, figs and other tropical and sub-tropical trees. By upper Cretaceous times most of the modern flowering plant lineages were established. Deciduous forests of beech, maple and plane trees were also common. Whole new groups of dinosaurs evolved after these changes took place, and presumably these late forms were adapted for survival in the new environments.

The low lying areas of the world surface provided a perfect medium for preservation of reptilian bones, but large tracts of high land must have been more truly terrestrial. We assume that the dinosaurs also invaded these areas when they established themselves as the dominant animals during the Jurassic. Dinosaur-containing rocks from the early part of this period are found in Europe, but as yet none have been found from the middle part of that time. This is probably due to the spreading of the seas which occurred throughout the middle Jurassic, leaving only truly terrestrial land masses with few swamplands; neither marine nor terrestrial environments were conducive to the preservation of dinosaur bones, although the marine deposits have been rich sources of information on fish, ichthyosaurs, plesiosaurs and the like. Dinosaur-bearing rocks of late Jurassic age contain a great variety of new forms indicating that dinosaur evolution must have been an ongoing process, presumably in the terrestrial uplands.

The Triassic was another period characterized by extensive continental hinterlands, but in contrast to the Jurassic uplands, terrestrial rock deposition took place and the Red Beds were formed, probably under rather arid conditions. In these beds are the remains of primitive reptiles called thecodonts, a group which showed some of the traits of their more famous dinosaur relatives.

A primitive thecodont such as *Ornithosuchus* looked superficially like a large lizard. It was about 1m (3ft) long but a good half of this length consisted of the tail. A closer inspection of the skeleton, however, would reveal several important differences between this animal and a lizard. The hind legs, for instance, were almost twice as long as the fore legs, suggesting that *Ornithosuchus* and many other thecodonts were bipedal; an interpretation which gains evidence from the structure of the hip. The pelvic girdle was very

Left: **This reconstruction of an early Cretaceous landscape shows *Megalosaurus*, a large carnosaur, in the middle ground. In the foreground is *Hypsilophodon*, a small primitive bird hip dinosaur, which may have lived in trees. Flying in the background is a pterosaur.**

securely attached to the vertebral column in the sacral region; immediately beneath this firm union the girdle, when viewed from the side, was developed as two processes, one pointing downwards and forwards, the other downwards and backwards. Both of these processes were very large and clearly served as anchorage points for large muscles of the upper leg. The shape of the upper leg bone or femur suggests that the hind legs of *Ornithosuchus* could be swung backwards and forwards directly beneath the body, rather like the legs of mammals. This type of leg movement is different from that shown by modern reptiles such as lizards, turtles or crocodiles. These modern reptiles show a typically sprawling gait where the legs are bent and held out from the body.

The bones were very lightweight and this fact, together with the stance, leads us to believe that *Ornithosuchus* was a fast, agile runner. When running it probably held the body horizontally, using its long tail to counterbalance the head and trunk. The fore limbs may well have been used to grasp prey, perhaps a reptile or insect, which was then dealt with by using the many sharp teeth in the jaws. These teeth were peculiar in that they were set within deep sockets in the jaws, a feature characteristic of a whole group of reptiles called the archosaurs, which includes not only the thecodonts but also the crocodiles, pterosaurs and dinosaurs.

There were two separate dinosaur lineages, the Saurischia or reptile hip dinosaurs and the Ornithischia or bird hip dinosaurs. It is doubtful if these two groups shared an immediate common ancestor; they probably evolved separately from an ancestral archosaur stock.

The first group, the saurischians, had made its appearance by the end of the Triassic. In many respects these dinosaurs were closely similar to their thecodont relatives; for instance many saurischians were bipedal, the fore limbs being developed for grasping, and some had extremely lightweight skeletons with hollow long bones. A technical point of similarity concerns the construction of the hip girdle described above; this gives rise to the name Saurischia – meaning reptile hip.

Basically there were three main groups of saurischians, coexisting for some 130 million years before dying out during the late Cretaceous. The coelurosaurs were one such group which were lightly built, fleet-footed carnivores. They were closely related to a second group, the carnosaurs, exemplified by that notorious villain, *Tyrannosaurus rex*. More distantly related were the sauropods, the giants such as *Brontosaurus* and *Diplodocus*. The skeletons of the coelurosaurs showed many of the adaptations of the thecodonts, a fact which suggests that they were running animals. The hind legs, however, were relatively longer than those of the thecodonts, since some

of the elongated ankle bones had fused with lower leg bones, a condition also found in modern birds. Indeed, the analogy with modern birds is even more striking when we consider some of the specialized coelurosaurs, the so called 'ostrich dinosaurs'. As their name implies, these dinosaurs were about the same size as modern ostriches, with large powerful legs, and a small head carried upon a long neck. There were three functional toes and the footprints left by these animals were remarkably bird-like.

Unlike true ostriches, however, the 'ostrich dinosaurs' possessed a long tail skeleton and the hand of the fore limb was used to grasp prey. One final feature worthy of note was the complete absence of teeth in the jaws – a fact that has led to the speculation that 'ostrich dinosaurs' had a horny beak, and that these creatures were egg eaters. It is worth mentioning at this point that although most dinosaurs were large, the coelurosaurs remained relatively small, most being 2.5–3m (8–10ft) in length. *Compsognathus*, a late Jurassic form, was diminutive by dinosaurian standards, being only about the size of a chicken.

The carnosaurs retained the primitive carnivorous habit, like the coelurosaurs, but in contrast to the small size of the latter the carnosaurs were very large. Everything about them was big and the longer the lineage lasted the bigger they became; the pinnacle of their evolution was reached by *Tyrannosaurus* and other Cretaceous carnosaurs such as *Gorgosaurus*. *Tyrannosaurus* was about 17m (55ft) long, stood nearly 6m (20ft) high and must have weighed around 8 tons. However, other carnosaurs were nowhere near as large, most, such as *Allosaurus*, reaching a mere 10m (37.8ft) in length. Bipedality was the rule among carnosaurs, and to carry all that weight the hind limbs were especially strong, with thick, solid leg bones and interlocking joints to provide strength. The foot was large with three large toes in front and one tiny toe behind which probably never touched the ground. The tracks left by these feet were remarkably like those of coelurosaurs and birds. One can therefore appreciate the initial puzzlement caused by the discovery of giant 'bird' footprints of Triassic age in the Connecticut valley.

The fore limbs of carnosaurs were very small, ridiculously so in *Tyrannosaurus*, and the fingers were reduced to three or even two in number. In most carnosaurs it is probable that the hand could be used for grasping and perhaps manipulation of food. The head was relatively larger than in any other dinosaur group, particularly in the later forms, and was supported upon a short, strongly constructed neck. Very long ribs upon the neck vertebrae suggest that the neck muscles were powerful, a necessity for an animal which tore at the flesh of its prey. There was a truly formidable array of dagger-like teeth, some reaching 15cm (6in) long in *Tyrannosaurus*, and along the

front and rear edges of each tooth there were keen serrated knife edges, somewhat like the edges of steak knives. This specialization must have greatly increased the efficiency of these teeth when dealing with flesh. If one examined the jaw of a carnosaur then one would see a great many teeth all of different sizes, large set alongside small; this feature is seen in all toothed reptiles, and demonstrates a complicated method of tooth replacement resulting in a constant supply of sharp teeth.

We can, of course, only surmise about the life style of *Tyrannosaurus* and its allies, but it seems likely that these giant carnosaurs were generally slow moving, and able only to achieve sudden bursts of speed over short distances. If we assume that the body activity was similar to that of a modern crocodile then, over long distances at speed, the bulky body of a carnosaur would be in danger of overheating. The concensus of opinion is that these creatures were active hunters, that they preyed upon herbivorous dinosaurs, but they may just as well have been scavengers, feasting upon chance carcasses rather like some modern hyaenas. For most people, *Tyrannosaurus*, *Allosaurus* and their relatives could have been nothing less than out and out hunters, truly living up to the word dinosaur.

So far we have considered dinosaurs as flesh eaters, insectivores, carrion feeders or perhaps even egg stealers. The sauropods however were primarily vegetarian. They were enormous, of dimensions never

Stegosaurus **was a Jurassic bird hip dinosaur, with a small head and a row of thermo-regulatory plates along its back. It fed on cycads and seed-ferns.**

to be repeated in the animal kingdom; for instance, the famous *Brontosaurus* was about 20m (60ft) long. Technically the sauropods were reptile hip dinosaurs, and like their carnivorous cousins, they had relatively small beginnings.

Plateosaurus was one such early (late Triassic) sauropod which reached about 6m (20ft) in length. It had many characteristics of its thecodont and coelurosaur contemporaries; for instance it was bipedal in stance with small front legs and large hind legs. The small head was supported on a long neck and the tail was long. However, in contrast to the coelurosaurs, *Plateosaurus* was a heavily built animal; its bones were solid and densely constructed. The difference in size between the fore and hind limbs was considerably less than in most coelurosaurs, the fore limb being almost two thirds the length of the hind limb and bearing four rather large toes. All this suggests that *Plateosaurus* went down on all fours from time to time, in which case it represented an early stage in the return to the quadrupedal locomotion that developed hand in hand with increasing bulk. The teeth were also markedly different from those of coelurosaur contemporaries. In *Plateosaurus* the jaw was beset with many peg-like teeth, completely unsuitable for flesh, but capable of coping with soft vegetation.

From these comparatively small beginnings true giantism became the rule, and by late Jurassic and Cretaceous times a number of forms reaching astonishing proportions were in existence. *Brontosaurus*, a late Jurassic sauropod, is well known. It was a typical giant reaching about 20m (65.6ft) in length and in life must have weighed around 35 tons. *Diplodocus* holds the record for greatest length at 29m (90ft), but was relatively slim – if we may use such an expression for so huge an animal. It probably weighed only 30 tons. Both of these types still showed their bipedal ancestry in having hind limbs slightly larger than the fore limbs. *Brachiosaurus* was even heavier than the other two with an estimated live weight of about 50 tons, but it was proportioned rather differently in that the fore limbs were longer than the hind limbs, a condition seen in the modern giraffes.

It is difficult to imagine why such giantism developed, for while enormous size confers certain advantages such as protection from predators, it also brings a host of problems. Indeed the structure of a sauropod dinosaur can best be interpreted in the light of the problem of size and weight bearing. Obviously throughout most of the sauropod skeleton the bones were massive. The vertebral column above the thorax and abdomen was particularly strong, with long neural spines on each vertebra for the attachment of powerful muscles and ligaments, which in turn supported the great mass of the viscera. The vertebrae were equipped with several processes which, when locked together with neighbouring processes, prevented up and down flexure of the vertebral column while still allowing a certain amount of side to side movement.

The sauropod pelvis was, as might be expected, a large, solid structure effecting a firm union between hind limbs and vertebral column. Much of the enormous weight was borne through this part of the skeleton, although the stance was quadrupedal, and the main locomotory thrust was directed through this point. The hind legs were constructed of massive, solid bones and acted as pillars, being held directly under the body; the weight, therefore, was directed through the axis of the leg. The front legs were similarly constructed, but the attachment to the backbone, via the shoulder girdle, was effected only by muscles and ligaments. This provided some degree of shock absorption during its ponderous movements. There may also have been a further form of shock absorption on all four feet, for there is a suggestion that behind the toes on each foot there was a large fleshy pad. Similar pads to these are to be found on elephants' feet.

Not all of the skeleton was solid and heavy. For instance, while the vertebrae of the neck were outwardly large, they were in reality designed for lightness and strength; at any position where bone was not needed for bracing or muscle attachment, a space appeared. This feature, found also in the skull, helped to reduce the enormous weight of the animal wherever possible.

It is almost certain that, in life, *Brontosaurus* and its allies were plant eaters, feeding on soft aquatic vegetation with weakly developed peg-like teeth. These teeth were often confined to the front half of the jaws. It is also highly probable that these giant sauropods spent a considerable part of their lives in water, where their

38

huge bodies could be buoyed up. The strongest piece of evidence for this supposition is the position of the nostrils, which were set very high on the head in *Brontosaurus*, sometimes above the level of the eye. This, by modern analogy, is an aquatic adaptation echoed in crocodiles. However like these present day animals the brontosaurs were not wholly aquatic; well defined tracks left as impressions in mud testified to the fact that journeys on to land were undertaken.

At this point in our survey of the dinosaurian world we may pause to consider a question of general relevance to the dinosaurs, but of particular relevance to the giants. The question is this: were dinosaurs warm blooded or cold blooded? Or, more precisely, could dinosaurs maintain and regulate their body temperatures internally, or did they use behavioural patterns to maintain a constant body temperature?

The maintenance of a constant internal body temperature is important for the thousands of chemical reactions inside the body. We know, for instance, the profound effects that result from a slight deviation from our normal body temperature of 37°C (98.6°F). Birds and mammals maintain their *temperature* internally. Heat is produced by breaking down food, and by activity; this heat is retained by an insulatory cover of fur or feathers and it may be lost by sweating or panting. Modern reptiles, on the other hand, while also having a remarkably constant body temperature, achieve this by their behaviour. During the early morning snakes and lizards bask, presenting as much

Left: Ornithosuchus, **a carnivorous thecodont, was one of the first reptiles to adopt a bipedal gait.**
Below: Triceratops, **a Cretaceous bird hip dinosaur, had a large head, three horns and a bony frill at the neck.**

of their body surface to the sun to absorb as much heat as possible. Later during the day they shun the searing heat and take refuge beneath rocks to avoid overheating. But could the dinosaurs also regulate their temperature by these behavioural means? Dinosaurs were, with few exceptions, large animals and it would take a very long time for a dinosaur to warm up or cool down. It is also difficult to imagine a dinosaur hiding beneath a rock! But, more importantly, a great deal of heat would be generated by long periods of strenuous activity, heat which would have to be lost by prolonged rest. Dinosaur skin, so far as we know, was impervious and dry like that of modern reptiles, so it was unlikely that they could sweat, and panting would not be sufficiently effective in lowering the temperature of a brontosaur. If we assume, therefore, that dinosaurs were like modern reptiles, this may have imposed restrictions on their activity. They may have moved slowly, or quickly but in short bursts. Of course we must remember that many of the largest dinosaurs spent most of their lives in water, a medium which equalizes out minor fluctuations in temperature.

There were one or two anatomical features which suggest that some dinosaurs at least were warm blooded. The microscopic bone structure of the brontosaurs and the bird hip dinosaurs was similar to that of mammals and, as this structure appears to be related to warm bloodedness, it is possible that dinosaurs had internal temperature regulation. Some of the dinosaurs, for instance the stegosaurs, had large plates along their backs. These plates, which were richly supplied with blood vessels, could have been used to lose excess heat. The large head frills of ceratopsian dinosaurs may have served a similar function, although

A model of Sir Richard Owen's reconstruction of *Iguanodon* can still be seen in a park at Sydenham, England. The reconstruction was rather inaccurate, especially with regard to the spike of bone which Owen placed at the tip of the nose instead of on the hand where it belonged.

it is clear that their main function was protection.

Warm bloodedness undoubtedly confers advantages. It means that the animal is, to a large extent, independent of its immediate environment, a particularly important attribute in higher latitudes. There are however disadvantages. Internal temperature regulation requires a great deal of food and it is questionable if the larger dinosaurs, such as *Diplodocus* and its allies, could eat enough food in a day to keep a warm blooded body going.

The second major dinosaur lineage was the Ornithischia or bird hip dinosaur, so named because of the construction of the hip which resembles that of birds (although it must be stressed that birds and ornithischian dinosaurs were but very distantly re-

lated). In ornithischians the pubic bone within the hip girdle had rotated rearwards so that it paralleled the ischium bone, and both projected downwards and backwards. Ornithischian bones are rarely found in Triassic or early Jurassic rocks; their heyday did not arrive until late Cretaceous times. However, we cannot pass entirely over the early forms, especially as one particular early Cretaceous form holds a special place in the annals of palaeontology.

In 1822 a young country physician, Gideon Mantell, was paying a house call in the Sussex countryside. Mantell had for many years shown a keen interest in local fossils, and on this particular occasion Mantell's wife, Mary Ann, accompanied him. While he was attending his patient she rummaged among the stones

as being very similar, although of course considerably larger than the teeth of iguanas. Mantell decided to name his teeth *Iguanodon* (iguana tooth), and he thus affirmed the reptilian nature of the extinct beast which bore them.

So it was that the first dinosaur was collected and properly described, although the actual word 'dinosaur' was not coined until some 20 years after Mantell's announcement of the existence of giant reptiles. *Iguanodon* has a further claim to fame however, for in 1834, a few years after the first teeth were found, a jumbled mass of bones was discovered in Kent and that allowed Mantell to associate his teeth with a partial skeleton. Later, more discoveries enabled specialists to piece together the skeleton of this ancient reptile. Sir Richard Owen, the author of the word 'dinosaur', restored *Iguanodon*. His restoration showed this animal to have been a thick bodied quadruped with short legs and a curious rhinoceros-like face, equipped with a horn on the tip of its nose, an unfortunately inaccurate restoration. A life sized concrete model was made under Owen's direction and set alongside other dinosaur restorations in a park at Sydenham, South London.

We now know quite a lot about *Iguanodon*, thanks to some keen-eyed miners of Bernissart, Belgium. In 1878 miners exposed a skeleton of a huge animal deep within a coal mine. With the help of staff from the Belgian natural history museum, they continued the excavations for several years and altogether recovered the skeletons of 31 individuals, a veritable herd, if 'herd' be an appropriate collective noun for a dinosaur gathering. But what these skeletons did show was an animal quite different from Owen's 'reptilian rhinoceros'.

Iguanodon was about 10m (32ft) long and was a distinctly bipedal animal, with large hind limbs, smaller fore limbs and a stout tail. It is probable that in life it rested on its hind legs and used its strong tail as the third 'leg' of a tripod. In this position it stood nearly 6m (20ft) tall. The 'horn', which was initially thought to have been attached to the skull, was in fact, a curious spiked thumb held upright in 'hitch-hikers' fashion. Presumably it was used for defence. The hind feet were not unlike those of carnosaurs, for there were three very large toes. *Iguanodon* was vegetarian, like all other ornithischian dinosaurs, and the skull showed specializations often seen in that group. Teeth were absent from the front of the jaws, which were developed instead as a short beak; however, there were a great many teeth within the mouth to afford an efficient grinding mill to deal with vegetation.

Iguanodon was a relatively primitive member of the ornithischian group, but other later bird hip dinosaurs developed unusual specializations. One such assemblage, the duck bill dinosaurs, were inhabitants of late Cretaceous times, and their remains are particularly

alongside the road; and so it was that Mary Ann Mantell found a few curiously shaped teeth which she showed to her husband. The teeth were broken but nevertheless were recognizably different from anything Mantell had ever seen before. Each tooth was leaf shaped with a narrow base and deep grooves on one side. The edges were finely serrated. Mantell showed the teeth to several eminent authorities of the day, all of whom professed them to be of little interest and none of whom suggested that they could have belonged to a reptile. Then a few years later Mantell, while in London, happened to meet Samuel Stutchbury. Stutchbury was a young naturalist who was then engaged upon a study of iguanid lizards from Central America. He immediately recognized the fossil teeth

common in the famous dinosaur graveyards of the Red Deer river valley of Alberta, Canada. They were generally rather similar to *Iguanodon*, except that the spiked thumb was absent; most were 10–13m (32–42ft) long and they must have weighed between 4 and 5 tons. The skull, however, was highly specialized. Teeth were confined to the back half of the jaws, and the front half of both upper and lower jaws was flattened from top to bottom and broadened from side to side. The appearance was similar to that of a duck's bill – hence their name. A duck bill such as *Anatosaurus* had a very large number of lozenge shaped teeth in its jaws, perhaps up to 1,000, and these teeth were being constantly replaced. The teeth were packed very tightly together and formed large batteries, the upper and lower series of which worked against one another as grinding mills. In addition, the lower jaw had large, upwardly directed processes which provided attachment for powerful chewing muscles. These dinosaurs evolved after a major expansion of the flowering plants had taken place; in contrast to the cycads and ferns, the leaves of flowering plants are hard and require considerable grinding before they can be digested. Presumably the batteries of teeth in the duck bill jaws were adapted to deal with the new vegetation.

Left: **An accurate restoration of the bird hip dinosaur,** *Iguanodon* **from late Jurassic times.**

Below: Anatosaurus **was an armoured bird hip dinosaur, about 4.5m (14ft) long with a broad flattened body covered by a mosaic of bony plates. It belonged to the ankylosaurs, a group sometimes known as 'reptilian tanks'.**

Two pieces of evidence suggest that these animals were amphibious. The tail was deep and flattened from side to side, like that of modern crocodiles, and was just the sort of tail which could be used for swimming. By some rare fortune of preservation, skeletons of *Anatosaurus* have been found in which webs of skin are discernable, stretched tightly between the finger bones. We may thus envisage the duck bills as feeding on the lush vegetation surrounding lakes, rivers and even coastlines, spending much of their time in shallow water and perhaps using deeper water as their refuge.

A form such as *Anatosaurus* was a relatively primitive duck bill dinosaur. There were, on the other hand, more specialized forms which developed surprising cranial protruberances. *Corythosaurus*, for instance, had a huge hollow helmet which ran over the back of the skull, and the skull roof of *Parasaurolophus* projected backwards over the neck as a thick tube which was almost twice the length of the skull itself. These cranial adornments were hollow and it appears that the nasal tubes wound through the structures before passing to the top of the throat. Understandably there has been much speculation about the function of these hollow crests with their greatly elongated nasal tubes. Some people have suggested that they provided the duck bill with a supply of air underwater, but the relative size of this supply was too small to have been of any real use. Another perhaps more plausible suggestion, is that the hollow crest acted as a resonating box, amplifying vocal signals produced in the throat. The shape of the crest may have affected the resultant tone. It may be that the crests were merely species

During late Cretaceous times duck bill dinosaurs were common. Many of them had curious, hollow head crests and all had batteries of strong teeth with which they could grind the leaves of the flowering plants and trees such as those shown in the illustration.

recognition signals and did not play any part in sound reproduction, but the notion of Cretaceous lagoons ringing with the 'nose flutes' of the duck bills is surely more appealing to the imagination.

Several other types of ornithischian dinosaurs abandoned the bipedal gait of iguanodonts and duck bills and reverted to a quadrupedal stance. One such group were the stegosaurs, exemplified by *Stegosaurus*, which lived in late Jurassic times – a cohabitant with *Brontosaurus* and *Allosaurus*. *Stegosaurus* was a bulky animal about 6m (20ft) long, and standing about 2m (6ft) high at the hips. This was the tallest point of the animal, due to the great length of the hind legs as compared with the small front legs. The tail was equipped with several spines and obviously served a defensive function, lashing out at attackers. *Stegosaurus* was famous for two reasons: the remarkably small size of the head and brain, and the series of large, bony plates borne along the back. The latter were huge triangular structures set in a paired row and, while their function is not definitely known, it seems possible

Brontosaurus was an enormous, solidly built sauropod. It probably spent part of its time in water where its vast body could be buoyed up and where it could feed on the soft vegetation at the water's edge.

that they played a role in temperature regulation, as explained previously.

The very small head of *Stegosaurus* was equipped with a few small teeth towards the back of the jaws, while the front of the jaws was produced as a horny beak. There is little doubt, therefore, that this animal was herbivorous, probably feeding upon the softer, lusher vegetation such as tree ferns. The truly remarkable feature of the head was the very small size of the braincase, which was about the size of a plum. On the other hand there was, at the base of the tail, a much bigger enlargement of the spinal cord. This has led to the popular belief that this and other dinosaurs had two brains. However, many reptiles have an enlargement of the spinal cord at the base of the tail, marking the position where many of the great nerves supplying the hind legs issue from the spinal cord. These nerves and their connections within the spinal cord are very important for the coordination of movement, obviously no less a problem for a dinosaur than for a crocodile. But this spinal enlargement cannot be considered as a brain, large though it may have been for the brain does far more than coordinate locomotion; it receives and sifts information from the environment and initiates reactions. Clearly the brain of *Stegosaurus*, although very small, must have carried out these functions.

What about the brains of other dinosaurs? Some calculations have been made by Professor E. Colbert who suggests that, on the whole, the dinosaurian brain was generally small. The brain of *Anatosaurus* would have weighed about 1/20,000 body weight; for *Brontosaurus* this figure would have been about 1/100,000, for *Tyrannosaurus* perhaps 1/10,000. This compares with similar calculations made for elephant and man, which are 1/1,000 and 1/60 respectively. However, brain size alone is a poor guide to estimated intelligence, for after all, a baby animal has a relatively larger brain than an adult. As Professor Colbert argues further, it is the various proportions of parts of the brain which can be far more instructive. The dinosaur brain, like the brain of other reptiles, showed large olfactory lobes and optic lobes, suggesting keen senses of smell and sight. The region of the brain associated with locomotory coordination, the cerebellum, was also well developed in contrast to the notably small cerebrum, the seat of intelligence. Clearly these animals acted by reflex actions rather than by intelligent learned behaviour as seen in mammals.

Another group of ornithischians, the ceratopsians or horned dinosaurs, were inhabitants of Cretaceous landscapes. Familiar examples of these dinosaurs were *Monoclonius* and *Triceratops*, both of which reached lengths of about 6m (20ft). Like other quadrupedal ornithischians the body was very bulky and the hind limbs were larger than the fore limbs, although only marginally so. Unlike *Stegosaurus*, the skull of a form such as *Triceratops* was relatively large, and the 'beak' was a very prominent structure, with the tip of the upper jaw turned down over the front of the lower jaw.

The most distinctive feature of the ceratopsians was the development of a large, bony frill from the back of the skull which overlay the neck, and clearly afforded protection to that most vulnerable portion of the anatomy. Some of these animals supplemented this protection with horns on the skull – a single horn in the case of *Monoclonius*, three horns in *Triceratops*.

One of the smallest forms, *Protoceratops*, gives us a rare insight into dinosaur biology. Remains of the 2m (6ft) long hornless adults were found in late Cretaceous rocks of Mongolia, together with remains of juvenile specimens and nests of eggs. There can be no doubt about which animal laid the eggs, for inside some of them were fossilized bones. Each of the eggs

Living in late Cretaceous times, *Ornithomimus* was about 2.5m (8ft) long, a lightly built coelurosaur of the type known as an 'ostrich dinosaur'.

was about 20cm (8in) long and about 10cm (4in) wide, long and narrow with a wrinkled surface. The largest nest contained 18 eggs, laid in concentric circles, although the outermost circle was unfinished, indicating that this was an incomplete nest. Since the discovery of the *Protoceratops* eggs many further finds have been reported and many previously unidentified eggs can tentatively be assigned to dinosaur origin. However, it is rarely possible to say to which particular dinosaur the eggs belonged.

Of course it is not surprising to discover dinosaur eggs, for the normal reptilian method of reproduction is to lay eggs; however, some reptiles, such as some lizards and snakes, bear live young. Is there any evidence that some dinosaurs did the same? The evidence is decidedly equivocal. The remains of two dinosaurs, *Coelophysis* and *Compsognathus*, have been found with smaller dinosaurs within the body cavity. Could these be unborn young? In the case of *Compsognathus* the contained bones cannot be identified with certainty, but with *Coelophysis* there is no question but that the included remains belong to the same species. The problem with this case, however, is that the presumed young are relatively large, and it is difficult to imagine how they could have been born. We must, therefore, reserve judgement on this question and introduce a more sinister interpretation. Perhaps these dinosaurs were cannabalistic on their young.

In sum, therefore, we must return to our original assumption that dinosaurs were egg layers. This fact has been used by several people to explain the most perplexing of all dinosaur problems – the question of dinosaur extinction. A few observations have suggested that towards the end of the dinosaur reign the shells of their eggs became thinner, perhaps in response to cooler temperatures and the onset of seasonality. Thinner egg shells imply greater mortality, which perhaps reached such serious proportions that extinction was the inevitable consequence.

There have, however, been a great many other explanations offered to explain the phenomenon of dinosaur extinction – some ridiculous, some facetious, others crude or rather subtle. It is as well to point out that, within the reptilian world, several marine forms such as ichthyosaurs, plesiosaurs and mosasaurs also became extinct, as did the flying pterosaurs. On the other hand crocodiles, snakes and lizards survived. So in seeking to explain dinosaur extinction we are looking for some selective agent. Immediately, therefore, we can dismiss many of the explanations which must have had universal effects. Some of these ideas are: earthquakes, magnetic reversal in the earth's crust, collision of the earth with a comet, meteorite showers, cosmic radiation, poisonous gases, increased radioactivity, volcanic dust, extraterrestrial invasion or worldwide floods. We can also reject several selective ideas because of their irrational basis. These include theories that dinosaurs became extinct because their bodies became too big for their brains, that they became racially sterile or simply tired of living, or that they became overspecialized.

Another idea, carrying more credence, suggests that dinosaurs were wiped out by disease – perhaps a virus partial to dinosaur physiology. The problem here is that one would have to assume that it became world wide in a relatively short space of time. Additionally it should be mentioned that pathogens very rarely destroy the hand that feeds them, and viruses are nearly always species specific. Competition from mammals has also been invoked, but the contemporary mammals were small shrew-like animals, and these would have had a very hard task to usurp the giants, devious though they may have been.

More acceptable ideas invoke indirect effects upon dinosaurs through the agencies of changes in vegetation or changes in climate. Clearly these two are closely

interrelated. There is indeed slight evidence that the climate was cooling throughout the Cretaceous period, but more significant was the development of marked seasonality. Continental drift was carrying huge land masses into higher latitudes, and the regime of distinct summers and winters favoured broad-leaved deciduous trees over the cycads and conifers. There is ample evidence suggesting that the flora underwent considerable change during the Cretaceous. Many of the more radical changes took place in the early part of the late Cretaceous, about 100 million years ago. Early relatives of the familiar oaks, maples and beeches made their appearance. However, for many of the herbivorous dinosaurs such as ceratopsians and stegosaurs, used to feeding on lush cycad leaves, the comparatively tough leaves and increased tannin content of these angiosperms may have caused feeding and digestion problems to which they were unable to adapt. This explanation does not explain the extinction of all dinosaurs, however, since the duck bills possessed a dentition which could surely have dealt with tough vegetation; so it is unlikely that the appearance of flowering plants caused problems for these animals. Additionally, most of the duck bill diversification took place after the appearance of the angiosperms.

The cooler periods of the late Cretaceous year may have caused problems for the larger herbivores and, indirectly, the carnivores. It is likely that, just as today, many of the flowering plants shed their leaves in winter, so removing a major source of food. Unless dinosaurs could migrate or hibernate (a difficult process for a large animal), as do many reptiles living today in northern climes, the result was inevitable.

Of course, the above mentioned speculations do not explain the extinction of the giant marine reptiles. We must also remember that some groups of dinosaurs, such as stegosaurs, had already become extinct, or were represented by very few types – for instance the sauropods had greatly diminished in numbers by Cretaceous times.

In conclusion, despite all the speculation, however carefully reasoned, we are left with no satisfactory explanation for the extinction of the 'House of Dinosaurs' which had ruled for 125 million years. Indeed, there is something to be said for not knowing the answer, thus leaving a certain amount of leeway for the fertile imagination.

The last word on this subject must surely belong to the American humorist W. Cuppy who, in his delightful book *How to become extinct* says: 'the Age of Reptiles ended because it had gone on long enough and it was all a mistake in the first place'.

Protoceratops is a small hornless ceratopsian from the Cretaceous rocks of Mongolia, where it is found in association with nests of eggs. Fossilized remains range from very small to adult specimens which reach about 2m (6.5ft) in length.

Chas. R. Knight
June 1902

Return to the Water

It seems unlikely that an 11-year old child could be responsible not only for the discovery but also for the excavation of a large vertebrate fossil. But a young girl did just this when she discovered the first articulated skeleton of an ichthyosaur (a fish-like reptile) in her local cliffs at Lyme Regis in Dorset, and arranged for its excavation. She was Mary Anning, born in 1799 and probably the first professional fossil hunter; she collected and sold her fossils, at first to visitors and tourists, later to museums and collectors from all over Europe. After establishing her reputation with her ichthyosaur she went on to collect the first articulated plesiosaur (near reptile) in 1824 and the first British pterosaur (winged reptile) in 1828.

At that time the relationships of these fossils were far from understood. Many eminent scientists of that time still believed in Special Creation and thought that the fossils found in the rocks had been drowned in the Great Flood of Noah or in a series of floods. Darwin's *The Origin of Species* was not published until 1859 and the controversy over his theory of evolution was intense. Today, thanks to our understanding of evolution we know that Mary Anning's ichthyosaurs and plesiosaurs were ancient marine reptiles.

For even as the earliest reptiles were evolving into true land animals some of them were returning to the water. This return has happened many times, not only within the reptiles, and representatives of many unrelated groups have become specialized for an aquatic existence. They have all faced the same problems – how to move in the water and how to obtain air. Although a variety of lifestyles was available to them, the majority became fish eaters and several times it happened that a later group adopted this lifestyle on the extinction of an earlier group.

It was during the middle Triassic that the first recognizable forms of crocodiles appeared and a rival group, the phytosaurs, appeared in late Triassic times. For a while phytosaurs flourished but by the end of

the Triassic they were extinct; apparently they could not compete with crocodiles. Phytosaurs were like large crocodiles with long flattened tails which were used for swimming. Their adaptation to life in water was incomplete for they still spent part of their lives on land and their limbs were used for walking, not swimming. Modern crocodiles spend a lot of time basking on mudflats but can lift the body off the ground and walk on all fours; it is assumed that phytosaurs could do the same.

Both the phytosaurs and crocodiles had elongated jaws with many sharp teeth – this sort of dentition is one we shall see repeated several times for it is perfectly adapted for catching fish. The position of the nostrils in the two groups of animals was different. In crocodiles, the nostrils are situated at the tip of the snout and are often on the top of a small bony projection; in phytosaurs they were located on the top of the head almost between the eyes. Crocodiles often lie in the water with just the top of the head and the tip of the snout projecting above the water; presumably phytosaurs did the same. In any event, the position of the nostrils would enable the animals to breathe while almost completely submerged.

The time of competition between the two groups was over by the end of the Triassic and since then crocodiles have been conservative, remaining essentially unchanged until the present day. The only major innovation has been the formation of a secondary palate in post-Cretaceous forms; this plate of bone separated the mouth from the nasal cavity thus ensuring that the mouth could be opened under water without interfering with breathing.

Most of the crocodiles inhabited shallow rivers and slow-moving waters but a few lived beside the sea. One group, the geosaurs, left the land completely and became adapted for a life in the open sea. The skin of these animals was scaleless and smooth, like that of a porpoise; the hind limbs were paddle-like and the tail was more like that of a fish's tail than a crocodile tail. The geosaurs were not a successful group, appearing and disappearing within the time span of the late Jurassic. It is highly likely that they could not

The ichthyosaurs were marine reptiles, mainly from Jurassic times. Shaped like a fish and with fin-like limbs, they were active predators, chasing fish and catching them with an array of sharp teeth.

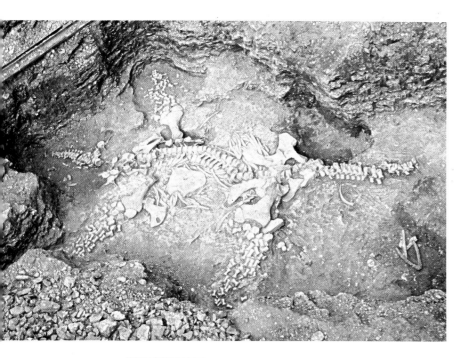

Above left: **The Fletton Plesiosaur. In 1970, plesiosaur bones were found in a brick-pit at Fletton, near Peterborough.**

Below: **Plesiosaurs were marine reptiles occurring throughout Jurassic and Cretaceous times. Many of them had long mobile necks, and all had paddle-like limbs with which they rowed through the water.**

Above right: **Many of the ichthyosaurs were beautifully preserved. Specimens from the Jurassic rocks of southern Germany show not only skeletal features but also the outline of the body, including the fleshy, boneless dorsal fin. Such comprehensive preservation is rare in the fossil record.**

compete with the ichthyosaurs and plesiosaurs which were already well adapted and successful marine forms.

It is difficult to believe that two groups of animals could start off with the same basic reptilian ancestry, invade the same marine environment and end up as different as the ichthyosaurs and the plesiosaurs.

The ichthyosaurs were reptiles shaped very much like the modern mammalian porpoises, sharing with those animals the absence of a neck which made the head blend into the body and resulting in almost perfect streamlining. The tail was fish-like and forked, but the vertebral column continued into the lower lobe of the fin whereas in most fish the column continues into the upper lobe. This sort of tail is called a reversed heterocercal tail and was found also in geosaurs. Ichthyosaurs were not very large, growing up to 3m (10ft) in length.

The plesiosaurs were reptiles with a very different appearance. They have at times been described as rowing boats with paddles but the most famous description was that by Dean Buckland of Oxford who described a plesiosaur as a 'snake strung through the body of a turtle'. They had relatively long necks (one branch of plesiosaurs, the elasmosaurs, developed necks which were up to twice the length of the body), short unarmoured turtle-like bodies and short tails. They were generally much larger than ichthyosaurs, growing up to 17m (55ft) long.

The limbs were adapted for locomotion in water in

both groups, but the process whereby this transformation was achieved was different in the two groups, and the functions of the fins or paddles were quite distinct. In ichthyosaurs the hind limbs and the pelvic girdle were greatly reduced, but the fore limbs formed large fins containing both extra digits (fingers) and extra phalanges (bones within each finger). The fins were used together with a large fleshy dorsal fin to balance and steer the animal, but the main propulsive thrust was provided by the body and the tail fin which undulated as in a shark. Ichthyosaurs could swim very fast and therefore could easily chase and catch fish.

The plesiosaur fore limbs and hind limbs were both transformed into large paddles (by the addition of extra phalanges only) and these were the main organs of propulsion, being used to row the animal along. Progress was probably relatively slow and these reptiles would have had difficulty in catching fish, were it not for their long necks, which were very mobile and which darted in amongst the shoals to reach the prey.

Certain features were common to both groups: both had external nostrils located far back on the top of the head to facilitate breathing while almost completely submerged; both had large eyes with which to spot shoals of fish; and both had numerous long sharp teeth for catching their prey.

A few ichthyosaur skeletons have been found to contain the skeletons of young. If they had not been eaten (and this possibility cannot be ruled out entirely)

then they must have represented embryos developing inside the mother when she died. These reptiles were highly adapted to a marine life and could not have moved on to land to lay eggs, as some people think the plesiosaurs did, so bearing young alive would have been distinctly advantageous.

At first sight then, the fish-like ichthyosaurs appear to have been much better adapted to their environment than the plesiosaurs. But both groups were present together throughout much of the Mesozoic, the ichthyosaurs appearing in the middle Triassic, the plesiosaurs in the late Triassic. They both thrived throughout the Jurassic but while the plesiosaurs survived to swim the seas of the late Cretaceous, surprisingly it was the ichthyosaurs which died out first. The famous Niobrara Chalk formation of the mid-western U.S.A. contains many plesiosaurs, but not one ichthyosaur has ever been found. Instead their place was taken by the mosasaurs, giant sea-going lizards.

The mosasaurs were found only in the late Cretaceous being especially common in the Niobrara chalk seas. They were 3–10m (10–32ft) long and had elongated heads and bodies with long tails which were used for swimming. The animals steered themselves through the water with large paddle-like limbs and caught fishes with numerous long sharp teeth.

The mosasaurs were successful for a while but at the end of the Cretaceous a wave of extinction swept the reptilian ranks and the mosasaurs did not escape. The

plesiosaurs also died out at this time and their places were left vacant awaiting the reinvasion of the seas by the marine mammals.

The group of mammals most truly adapted to a life in the sea is the cetaceans – the whales, dolphins and porpoises. The first members of the group appeared in the Eocene and were already well adapted to life in the water. Their bodies were elongated and the tail was transformed into a horizontal fluke. Swimming was accomplished by undulatory movements of both body and tail. The hind limbs were already vestigial while the fore limbs were large paddles used for steering. They varied in size from that of a porpoise to real giants such as *Basilosaurus* which reached up to 27m (85ft) in length.

These primitive whales did not have true blowholes since the external nostrils were only half-way back along the skull, unlike the condition in modern cetaceans, where the external nostrils form a fully developed blowhole at the top of the skull. They had long jaws with numerous sharp teeth which were used for catching fish. The fish-eating habit has persisted in many of the more recent whales, and in dolphins and porpoises. However the very largest of the cetaceans, the whalebone whales, became modified for catching plankton; at the back of the mouth is a fringed network of whalebone (hardened skin) and the animals swim slowly along using the framework to strain the plankton from the water. The small animals

Above: This modern crocodile is similar to the early forms which lived in Triassic times. Like them, it is adapted to a semi-aquatic existence.

Right: Basilosaurus was a primitive Eocene whale in which the nostrils had migrated halfway along the snout. (A major trend in the evolution of the whales was the migration of the nostrils or blowhole from the tip of the snout to the top of the skull).

are then licked from the whalebone with the tongue and swallowed.

The whalebone whales are the largest animals that have ever lived on this planet, the blue whale weighing as much as 150 tons. They share the planktonic habit with the largest of the sharks; it seems that giantism and fish-chasing do not go together. Once the animals reach a certain size they cannot afford the energy expenditure needed to chase fish, but must be content to drift along straining out the plankton.

The marine reptiles and mammals are amongst the most fascinating animals of our modern world. But the great whales are in danger of extinction, seal numbers and habitats decrease every year and turtle eggs are considered a rare delicacy as is turtle soup. As recently as the 18th century Steller's sea-cow became extinct, leaving only two genera of this group remaining. If the current trends continue it may be that in a few hundred years the seas will be left again to the invertebrates and to the fish. Ours will be the loss.

Life in the Air

In 1959 an industrialist, Henry Kremer, offered a prize of £5,000 (later raised to £50,000) to the first person to fly a mile-long figure of eight course, beginning and ending with a hop over a 3m (10ft) high wire. It was stipulated that muscle-power alone was to be used; no engine of any kind was allowed. But the task was to prove a difficult one indeed and it was not until 1977, after numerous abortive attempts by many people, that the flight was made and the prize secured.

It is probable that the history of natural flight followed a similarly experimental pattern. We do not know how many abortive attempts have been made to develop flight since life originated on this planet over three thousand million years ago. The fossil record of flying organisms is inevitably poor, since these animals had light and fragile bodies which quickly disintegrated after death; only rarely would they have fallen

Left: A 'megadragonfly' drifts over a Carboniferous swamp, where ferns and giant clubmosses line the water's edge.
Below: An insect in amber.

into mud, or some other suitable preserving medium. To our knowledge only four groups of animals have mastered true flight, namely insects, birds, bats and the extinct pterosaurs. This rarity of flying organisms is due in great part to the difficulty of life in the air. Air is thin and gives little buoyancy, unlike water; therefore an airborne animal must have efficient bracing for the softer parts of the body, in contrast to a water-dwelling animal which can depend on the water for its support.

Major problems of flight are those of take-off and landing. In the first case, enough lift must be gained to raise the animal off the ground, and in the second case, some way of cushioning the impact of landing must be found. Some flying animals have solved these problems more efficiently than others. Albatrosses need a runway several feet long at the top of a cliff to enable them to get airborne; bees have vertical lift-off. Puffins land on water with what can only be described as a splash-down while a fly can land on a ceiling, turning upside down as it does so.

Insects are in many ways the most highly adapted of all flying organisms. Their bodies are supported by a chitinous exoskeleton and delicate projections from this exoskeleton form the wings. They were the first animals to take to the air, primitive wingless forms being present in the early Devonian, and winged insects being found in the Carboniferous. That age witnessed the beginnings of an evolutionary line that was to continue to the present day. Winged insects are probably the most successful group of animals this world has ever known. However, their fossil record is poor, as is so often the case with flying organisms, but the few that have been found are of great interest. The largest known insects were the 'megadragonflies' which flew over the Carboniferous swamps. They had large, gauzy wings with spans of up to 60cm (24in) although their bodies were probably about the size of a modern dragonfly body. Many insects have been found embedded in amber (fossil resin). Presumably the animals were trapped in the resin as it was exuded from the bark of coniferous trees, and then the insects and the resin were fossilized together.

Insects have survived the vicissitudes of climate and vegetation that have occurred since the group first took to the air, as have birds and bats. But a fourth group of flying animals, the pterosaurs, were less fortunate, living only during the Jurassic and Cretaceous. There were two groups of these flying reptiles, the earlier having long tails while the later and more specialized forms had shortened tails.

Rhamphorhynchus was typical of the early forms, living during middle Jurassic times. It was about 75 cm (30in) long and had a long, mobile neck, long tail and a shortened body. The jaws were elongated and beset with numerous sharp teeth usually seen in fish-eating animals. However, these teeth were unique in that they sloped forwards instead of backwards as in most animals. The tail was also unique in this animal, since it was flattened at the far end and may have acted as a rudder during flight.

Although the later forms still had the long neck and shortened body, their tails were considerably reduced (as are those of the unrelated modern birds). Several evolutionary trends can be seen in this later group such as increase in size, loss of teeth and the development of a backwardly directed crest on the skull. In many ways *Pteranodon* represented the end point of these trends, with a wing span of up to 8m (26ft), no teeth and a crest which was nearly as long as the main

part of the skull. Remains of this creature are found in the late Cretaceous Niobrara Chalk formation of Kansas, a deposit thought to have been laid down in a shallow nearshore sea.

A large *Pteranodon* in flight must have been a spectacular sight, wheeling above the waves. Each wing consisted of a web of skin stretched tightly on the bones of the forelimb and the greatly lengthened fourth finger. However this structure had inherent deficiencies. There were no internal strengthening struts and any tear in the wing would seriously impair the flying ability of its owner. In contrast, a bat's wing, while also formed by a web of skin, has four bony fingers penetrating its structure imparting greater intrinsic strength. A tear in any one section does not destroy the whole wing.

It has been suggested that pterosaurs were incapable of fast flight. If this were true then they could only survive in conditions of light air currents such as probably occurred throughout much of Jurassic and Cretaceous times. Accordingly it is likely that most of their flight consisted of cliff-top soaring and slow gliding down to sea level; flapping flight and up-currents would have been used to reach the cliff-top again. The pterosaurs may have landed on the sea while feeding; if so it has been suggested that they might have perched on wave tops during take-off to avoid entangling their wings in the water. Alternatively, they may have caught fish on the wing and so avoided the need for landing on the water.

The hind limbs of a pterosaur were weak structures, probably incapable of supporting the weight of their owner. They may have been used, like a bat's hind limbs, for hanging upside-down on craggy cliff-faces. Landing must have been a hazardous process and it is thought that the animal alighted on its feet on the cliff-top and then flopped forward. It could then partly drag, partly push itself along to its hanging position against the cliff-face. Take-off would then be relatively easy, the air currents providing lift to the outstretched wings as the animal pushed itself off the cliff.

The pterosaurs became extinct at the end of the Cretaceous. Speculation as to the causes have included unsuccessful competition with birds which were becoming an important element in the fauna, and a

Above: Hesperornis **was a large flightless bird from Cretaceous times. Its wing bones were very reduced and although it is shown here with wings, experts differ in their opinions as to exactly how much wing was visible.** *Below: Ichthyornis* **was a small flying bird with well developed wings and an obvious keel on the sternum. (These two birds are not drawn to the same scale).**

change in climatic conditions. If, as has been suggested, the pterosaurs could fly only in light winds, then an increase in wind speed could well have resulted in their extinction. Wind speeds are controlled by temperature differentials between the poles and the equator, and during most of the Cretaceous this was only 24°C (77°F) compared to the present day 48°C (188°F). If, at the end of the Cretaceous, an increase in the temperature differential (and there is some evidence for this) led to an increase in wind speed, then the birds may have been able to cope with the changed conditions better than the pterosaurs.

Like the pterosaurs the birds are related to the dinosaurs. In fact, unlikely though it may seem, many experts believe that birds are very close relatives of the dinosaurs. Much of the evidence for this relationship lies in *Archaeopteryx*, the first and most primitive of the birds. The original specimen of this animal (its name means ancient wing) was found in a lithographic limestone quarry in Bavaria in 1860, and it caused great excitement amongst the workmen.

The body form of *Archaeopteryx* was similar to that of a small running dinosaur, and other reptilian features included a long, lizard-like tail, teeth, and claws on the fore limbs. However, these same fore limbs were partially modified as wings with clear impressions of feathers on the fine-grained limestone in which the fossil was preserved. Other bird-like features included a wishbone and a big toe that was opposed

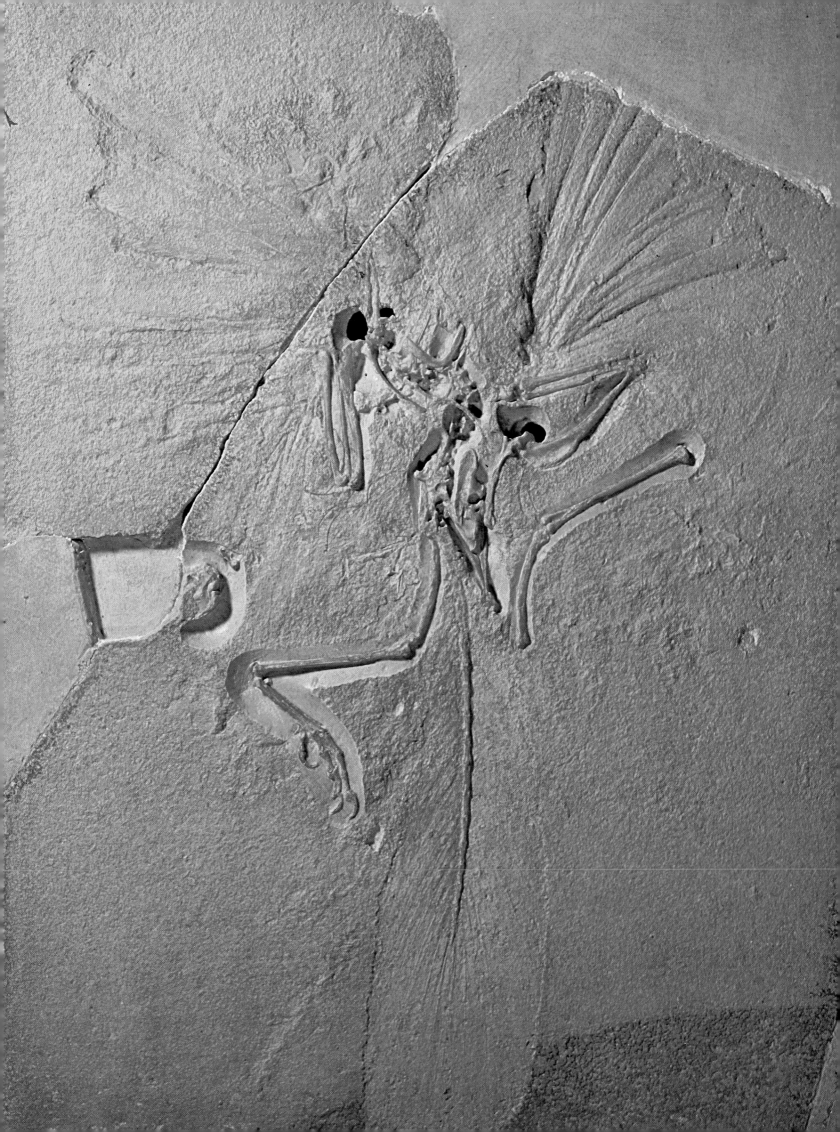

Left: The first articulated specimen of *Archaeopteryx lithographica* was found in Bavaria in 1862. It shows quite clearly the imprints of the feathers which stamp this animal as a bird despite its many reptilian features.

Above: This reconstruction of *Archaeopteryx* shows the long tail and the teeth, both features indicative of the reptilian ancestry of this primitive bird.

to the other three toes. *Archaeopteryx* was about the size of a magpie.

There has been much controversy over the life style of this animal. Some experts think that it led a running life, in which case the wings may have functioned as insect traps, rather like large butterfly nets, with the feathers helping to increase the surface area. Alternatively, the wings may have acted as aerofoil planes, so increasing the running speed. Another theory proposes that this primitive bird spent its life in trees. The claws on the wings would have facilitated scrambling among the branches, and the opposable toes certainly seem better adapted for perching rather than for running. In either case true flight seems unlikely. There was no keel on the sternum such as is found in most modern birds, so the development of flight muscles must have been poor. Feathers probably originated as insulatory structures to maintain body heat and only later became modified for their use in flight.

The subsequent history of birds suffers from the paucity of its fossil record. *Archaeopteryx* lived in late Jurassic times and fossilization in the quiet lagoons which formed the lithographic limestone was indeed

a fortunate accident, since only in such an environment could the fragile skeleton have been preserved. It is not until the late Cretaceous that more bird remains give us another glimpse into avian history, and it is in another fine-grained rock, the Niobrara Chalk, that they are found.

Hesperornis (the name means western bird) was a marine fish-eating bird that in many ways was a strange mixture of primitive and advanced features; for example it still had teeth, but the tail was considerably shortened. This Cretaceous bird was peculiar in that the fore limb and the bones of the shoulder girdle were very reduced, and it is likely that little or no wing was visible externally. The hind feet were webbed for swimming. *Hesperornis* probably led a diving life similar to that of a modern loon.

A contemporary of this bird was *Ichthyornis* (meaning fish bird) which led a very different kind of life. It was obviously a strong flier since the sternum was provided with a distinct keel for the attachment of powerful flight muscles, and the fore limbs were modified into true wings. It was about 30cm (12in) long in contrast to *Hesperornis* which was huge, about 2m (6ft) from the tip of its beak to the tip of its tail.

Other birds have been found in Cretaceous rocks, all but one from marine deposits. This pattern of fossilization continued into the Tertiary and consequently our knowledge of terrestrial birds is negligible. Since the number and variety of aquatic birds increased considerably during the Eocene epoch, we can only assume that a similar increase in the terrestrial forms also took place.

The vast majority of Tertiary birds had typical avian bodies with light tubular bones, flexible necks, shortened trunks, and tails formed solely of feathers. Their skulls were large in comparison to those of their forebears and their jaws were developed as toothless beaks.

Some birds, however, became modified for unusual modes of life. The penguins, for instance, had already adopted a marine existence by the Eocene and had succeeded in invading a highly inimical environment in their life on the southern polar ice cap. Their wings became modified to form powerful flippers which could be used for 'swimming', sometimes described as flying underwater. Surprisingly, the early penguins were quite large, standing over 1m (5ft) in height, and one of the major trends in their evolution has been a reduction in height. The modern Cook Strait Blue Penguin for instance, stands a mere 40cm (16in).

A greater contrast than that between the penguins and *Diatryma* is difficult to imagine. *Diatryma* was an Eocene ground bird, a giant in avian terms, standing over 2.5m (6ft) high, with large, heavy leg bones and a heavy skull armed with a powerful beak. However, the wings were much reduced and it was obviously incapable of flight. It has been described as 'one of the landmarks of avian palaeontology'. Presumably it had few enemies since the mammals of that time were small and the giant reptiles had long been extinct.

Since the middle of the Tertiary a series of other gigantic, flightless birds has appeared, but these later forms were different from *Diatryma* and all other birds in a whole suite of characteristics, particularly connected with the skull and the breastbone. These ratites (so-called because the breastbone is without a keel, 'raft-like') include the modern ostriches, emus,

Pteranodon **was one of the largest of the Cretaceous pterosaurs with a wingspan of up to 8m (26ft). It used warm air thermals to sustain its soaring flight.**

kiwis and the like, but probably the most fascinating of them all was *Aepyornis* of Pleistocene age. It was over 3m (10ft) high and may well have inspired the legends of the roc of Sinbad the Sailor, living as it did on the island of Madagascar. Eggs of this remarkable bird were up to 75cm (30in) in girth and probably contained two gallons of liquid; they are still often exposed by the erosion of the sandy soils in which they were laid all those thousands of years ago, and some even contain the bones of once developing chicks.

The activity of flying demands a great expenditure of energy. Pterosaurs lived at a time when climates were generally warm and the impression of hair-like projections on the rocks around their bones suggests that they may have had some insulation of sorts. Birds, of course, are warm blooded and maintain a high body temperature aided by their extremely efficient insulatory feathers. Bats, although their bodies are furry, have large expanses of bare skin on their wings and can only maintain a high body temperature in flight. At rest their body temperature soon drops to that of the surrounding air and they survive cold winters by going into deep hibernation. Insects cannot maintain high body temperatures and in the face of winter cold they soon die. Their eggs and pupae can survive the rigours of inclement weather until spring returns.

Considering all of the difficulties faced by a flying animal it is surprising that any managed to invade the air. However, when we look at the many ingenious adaptations evolved as a response to the difficulties, then it is not surprising that amongst those organisms which did succeed are found the most numerous and successful of all animals.

Diatryma was a giant flightless bird which lived in North America and Europe during early Eocene times.

The Age of Mammals

It seems strange to consider that, even as the small dinosaurs of the late Triassic age were heralding the advent of their numerous and famous reptilian descendents, the mammals were already roaming the earth. However, unlike the dinosaurs which dominated the landscapes for the rest of the Mesozoic, the mammals remained small and hidden for the whole of this time until the dinosaurs became extinct at the end of the Cretaceous. Then, and only then, did the mammals begin an explosive period of evolution that ended in their occupation of many of the roles of the extinct reptiles. It is worth remembering that the 'Age of Dinosaurs' lasted for over 125 million years, more than twice the time of the Tertiary 'Age of Mammals', and that the Mesozoic mammals were present for about 90 per cent of that time.

They survived by maintaining a low profile, ranging in size from that of a mouse to that of a cat. They may have been arboreal, possibly nocturnal, and they were almost certainly secretive, like modern shrews. Their basic diet was probably insects, although some were herbivorous rather like the rodents of today. Most of them became extinct before the middle of the Cretaceous and very few survived into the Tertiary. It is possible that the egg-laying mammals of Recent times (the platypus and the spiny anteater, both from Australasia) are relics of this vanished array of Mesozoic mammals.

By the end of the Cretaceous, representatives of the two major modern groups of mammals had appeared in the fossil record. It is possible that both groups are derived from the same Mesozoic mammals – they had a common ancestry. They are the marsupials, (pouched mammals such as the kangaroo) and the placentals (mammals that bear their young until they are more or less independent such as the dog). The marsupials are now found in Australasia, South America and North America, while placentals are found all over the world except Australia (the few placentals that live in Australia have been introduced there by man during the last few centuries, except for a few bats which flew there by themselves). Unfortunately, as yet, no mammal fossils have been found in Australia before late Tertiary, so that the mammalian history of that continent remains a mystery.

The discontinuous distribution of modern mammals has been the subject of much controversy. We now think that during the Cretaceous period there were

Left: Megaloceras **(the Irish Elk) was really a giant deer from the Pleistocene epoch. Shown here is the male which carried enormous antlers on its head; the female had none.**

Right: Alticamelus **was another mammalian giant, standing over 3m (10ft) tall on long stilt-like legs. These large Pliocene camels are sometimes called 'giraffe camels' for obvious reasons.**

Eohippus, the 'dawn horse', had moderately long legs with four functional toes on the fore feet and three on the hind feet. These horses were therefore not adapted for running as were the later grassland forms, but lived in woodland areas, where they browsed on leaves and soft vegetation.

two large continents, Laurasia in the northern hemisphere and Gondwanaland in the southern hemisphere. (At least the southern American portion of Gondwanaland contained both early marsupials and early placentals.) By early Tertiary times Gondwanaland had split into separate South American, African, Indian and Antarctica/Australian portions. It is clear that marsupials alone were present in the Australian portion while a mixed fauna of early placentals and early marsupials were isolated in the South American portion, and these animals continued to evolve in isolation from the rest of the world until the Pleistocene epoch. As yet, only placental fossils have been found in Africa and India, and the marsupial history of those continents (if any) is still obscure.

There was probably a mixed fauna of marsupials and placentals in Laurasia (North America, Europe and Asia) at the beginning of the Tertiary, but then the marsupials rapidly died out, while the placentals underwent a period of rapid evolution. During the Miocene, Africa and India collided with Laurasia and then the faunas of these continents mixed. In the Pleistocene a connection between South America and North America was formed with profound consequences for the South American fauna.

The fauna of the South American continent, up to the Pleistocene epoch, was an unfamiliar one to our modern eyes. There were marsupial carnivores and rodent-like forms, while the placentals were mainly herbivorous. Among the strangest of the herbivores were the glyptodonts, relatives of the modern armadillos; they could be described as the armoured tanks of the mammal world. The armour took the form of a massive dome of bony plates which covered the body, while further plates protected the head. Some of these animals even had a spiked knob of bone, like a medieval mace, at the end of an armoured tail – enough to deter any prospective predator. As can be imagined, the legs were very large and heavy, to carry the enormous weight of the body. A hungry carnivore would have had a problem knowing where to attack, even if it survived the onslaught of the tail.

Huge relatives of the glyptodonts were the giant ground sloths. They were not armoured, although they did have bony plates in the skin. Some of them were as large as elephants, with heavy skeletons to match. They must have looked very clumsy since they walked on the sides of their hind feet and on the knuckles of the front feet. Their relatives, the tree sloths, still inhabit the jungles of South America and they are still moving with a slow and clumsy gait.

A remarkable array of strange South American ungulates prospered throughout the Tertiary. One form, *Toxodon*, was first discovered by Charles Darwin while he was on his world tour in the *Beagle*. He took the skeleton back to Britain with him, where it was described by the eminent palaeontologist Sir Richard Owen. *Toxodon* looked like a 3m (10ft) high cross between a hippopotamus and a large rodent with short front legs. Many South American mammals showed parallels with modern mammals; for instance

Glyptodon **was a Pleistocene glyptodont, a group which appeared in the Eocene and became extinct at the end of the Pleistocene. It was nearly 3m (10ft) long and had the armoured body of its kind.**

Thoatherium, an extinct ungulate, exhibited many similarities to modern horses, presumably because both animals lived in similiar environments. *Thoatherium* and the modern horse are the only animals ever to have achieved a one-toed condition. Their limbs became elongated to facilitate fast running and only the central toes remained functional as hooves, after the streamlining of the limb.

Yet another South American mammal, *Pyrotherium*, paralleled the modern elephant in its evolution. It was a gigantic animal with tusks and probably a trunk. Its life style was probably similar to that of modern elephants, pushing through bush and forests and browsing on the upper leaves as it went. All of these South American herbivores were preyed upon by marsupial carnivores. Again, parallels can be drawn between them and placental forms: *Borhyaena* was a wolf-like animal; another, *Thylacosmilus*, resembled a sabre-toothed cat.

The unfamiliar mixture of mammals came to an end during Pleistocene times, when the Panama isthmus came into being. Then there came an influx of placental sabre-toothed cats, lions, wild dogs and others. These placental carnivores were probably faster and more cunning than the marsupial forms, and the marsupials could not compete. Large numbers of modern horses, deer and llamas also came, and were probably more efficient browsers and grazers than the existing herbivore population. Under the onslaught of carnivores and competitors the existing ungulates died out.

Marsupials and placentals probably lived together in North America, Europe and Asia during the late Cretaceous. But at the beginning of the Tertiary a rapid evolution of placentals took place, and the marsupials, if present, rapidly became extinct. Since the brain cases of the placental mammals were generally larger than those of the corresponding marsupials, it is usually assumed that the placentals succeeded where the marsupials failed because they were more intelligent. Marsupials also differed in that they had no milk teeth, and their teeth were often rather unspecialized, in contrast to the highly specialized dentitions of the placentals. Efficient teeth made for efficient feeding and this may well have contributed to placental success. The earliest placentals were small, shrew-like animals, and about 28 orders of placentals have evolved from this basal stock, of which 16 groups are still in evidence. Much of the basic differentiation of the mammal groups took place very early in the Tertiary.

The ungulates – large, hoofed, herbivorous mammals – appeared very early, and some of the early forms were certainly bizarre in appearance. *Uintatherium* lived in North America and Asia and was about the size of a large rhinoceros. It had thick limbs in which the upper bones were long, while the lower bones and ankles were short; the feet were broad and spreading. The animal was provided with no less than six horns on its head, four of which pointed in different directions, and it had large protruding canines. This grotesque ungulate had a small brain, and for all its size, horns and canines, it became extinct by the end

This small Mesozoic (late Triassic) mammal, *Eozostrodon*, **still showed its reptilian ancestry in some features of its braincase and jaw structure. However it had mammal-like teeth and a hairy skin.**

of the Eocene epoch. *Uintatherium* was one of a whole series of slow, early ungulates with short evolutionary lives. By the middle of the Oligocene they had been superseded by the modern ungulates, which were faster and had larger brains than the early forms.

The odd-toed ungulates are represented in modern faunas by horses, tapirs and rhinoceroses. One of the most interesting things about the evolution of the horse is the extraordinary completeness of its fossil record. The sediments from North America in which the fossil horses are buried are complete from early Eocene to Recent times, and from early Miocene onwards were characteristic of a savannah situation. *Eohippus*, the dawn horse, was small (about the size of a fox) with a small head and low-crowned teeth. It probably browsed on soft vegetation and leaves. Through the evolution of the horses several trends can be discerned: firstly, the horses increased in size until today they stand up to 2m (6ft) high. Most of this increase in height is accounted for by a tremendous elongation of the limbs, especially in the bones of the foot which have become part of the leg. Only the hoofed middle toe has survived this elongation; all the other toes gradually became reduced in size during the long course of evolution. This change in stature is associated with adaptation to a running mode of life.

The other major evolutionary change which characterizes the horses is a change from a browsing to a grazing mode of life. This change is probably associated with the spread of the grassy plains during the middle of the Tertiary. Grass is very hard and tough, and the horses gradually evolved high-crowned, complex teeth to deal with the grass, which needs a lot of grinding before it can be digested. The teeth are worn down very quickly by the tough grass, and the complexity of the horse's teeth ensures that an efficient grinding surface is maintained. To accommodate the larger teeth, the jaws became longer and deeper. This, together with the enlargement of the brain case, accounts for the evolution of the larger head of modern horses.

The irony of horse evolution was that they became extinct in North America, where the whole of their evolution had taken place. The partnership which grew up between early man and horses in Europe and Asia was only possible because some horses had migrated across the Bering land bridge before the North American horses became extinct in the Pleistocene.

Later, horses were reintroduced into North America, where they are so much a feature of the mid-west.

One group of odd-toed ungulates had a brief but spectacular evolutionary life. These were the titanotheres, which appeared in the early Eocene, rapidly became successful gigantic beasts emulating the rhinoceroses, and then just as rapidly died out in mid-Oligocene. One of the most impressive was *Brontotherium*. It stood about 2–4m (6–12ft) high and looked like a cross between a rhinoceros and a buffalo, with a strange cleaved horn on the front of its head. It had the usual adaptations to a heavy build, massive limbs and a strong, heavy skeleton. The titanotheres were amongst the most abundant fossils found from the mid-Oligocene of North America, yet just after this time they became extinct. Since their teeth were adapted for browsing on soft vegetation and could not grind the tough grasses, it is probable that the spread of the grasslands at that time contributed to their extinction.

The chalicotheres would have been considered unremarkable odd-toed ungulates, but for one very unexpected feature. They lived throughout Tertiary times, looked rather like horses, had teeth like titanotheres – but instead of hooves they had clawed feet! When the first feet were discovered in 1823, Cuvier (an eminent anatomist of his time) claimed that they belonged to an anteater. It was only when the feet were found attached to a complete fossil skeleton that the truth became known. Many speculations as to the life and habits of the chalicotheres have been made. Some people think that the claws were indicative of a swamp life where they could be used to dig up roots and tubers. Others think that these animals lived in forests and used the claws to cling to trees or to drag down branches while they browsed on the leaves. Their skeletons are often found with those of other known forest dwellers, a fact which supports the latter theory.

Our final group of odd-toed ungulates, the rhinoceroses, are probably upon the brink of extinction. Having maintained their numbers throughout the Tertiary, there are now only a few species left. The earliest rhinoceroses were found in the Eocene, and they looked much more like horses than do their modern counterparts. Sometimes known as 'running rhinoceroses', they were quite small and lightly built with long legs that were well adapted for a running mode of life. Most of these early forms were hornless, and this feature was equally true of *Baluchitherium*. No greater contrast can be imagined than that which existed between the running rhinoceroses and the giant *Baluchitherium*, for this latter animal claims the distinction of being the largest land mammal known. It was about 4.5–5m (15–17ft) high and about 7m (23 ft) long, and lived in Asia in Oligocene times, where it probably browsed on trees in much the same way as the modern giraffes.

The even-toed ungulates comprise the vast majority of modern ungulate forms, including the pigs, hippopotamuses, camels, giraffes, deer, antelope, sheep, goats and cattle. Their fossil record is as varied as their Recent history, so it will be possible to review only a few representative types.

Many of the early even-toed ungulates resembled modern wild pigs. One of these, *Dinohyus*, was not quite as grotesque as a wart-hog, but it had a unique ugliness all its own. About the size of a bison, its head was enlarged quite out of proportion to the rest of the body, being about 90cm (3ft) long. The portion which housed the brain was still quite small, however, most of the increase in size being due to the elongation of the face. These early pig-like animals survived only into early Miocene, since they were apparently not intelligent enough to compete with the true pigs which were evolving rapidly at that time.

Camels form another even-toed ungulate group. Their history was mostly centred in North America; and then, like the horses, they became extinct in the Pleistocene. They survived because some camels migrated to Europe across the Bering land bridge before extinction overtook the indigenous population. The related llamas migrated south across the Panama isthmus to survive in South America. An important trend seen in camels, but unknown in other even-toed ungulate groups was the loss of hooves. Eocene camels had hooves, but gradually, over the millenia, these were lost and were replaced by nails and large pads. These modified feet contribute greatly to the ease with which camels walk over sand, an attribute which has gained for them the name 'ships of the desert'. One Pliocene form, *Alticamelus*, was over 3m (10ft) tall with extra long legs and a long neck like a giraffe. It probably lived in bush rather than in desert, browsing on the tops of the trees and running away from predators with its long legs.

If the giraffes had become extinct millions of years ago we would probably have considered them very strange animals. In fact the fossil forms are all very much less remarkable than the Recent representatives, being short-necked forest dwellers, much like the modern okapi. The sivatheres from the Pliocene and Pleistocene of India had enormous bony horns on their heads and they were much stockier in build than many of the other giraffes. It is quite possible that some of the early men may have spotted this unusual giraffe moving slowly through the trees and browsing on the leaves.

Perhaps one of the most spectacular of the extinct

Distantly related to the elephants were the dinotheres which lived from the Miocene to Pleistocene times. This species of *Deinotherium* was about 3m (10ft) high and, like all its relatives, had two large down-curving tusks projecting from the lower jaw.

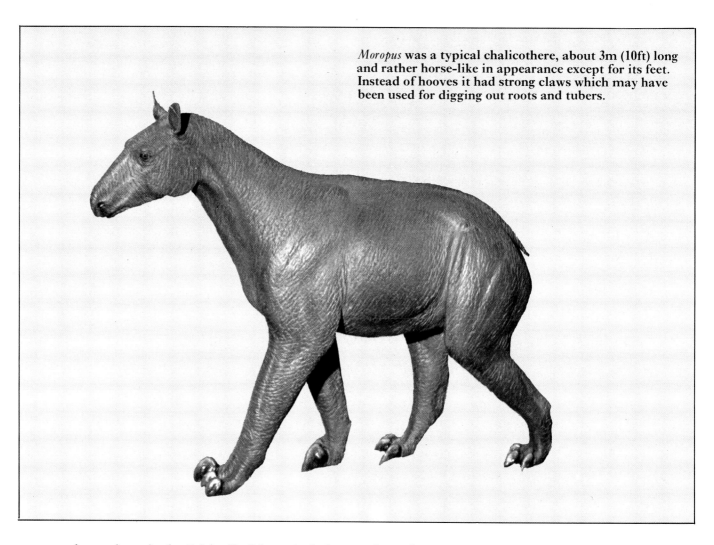

Moropus was a typical chalicothere, about 3m (10ft) long and rather horse-like in appearance except for its feet. Instead of hooves it had strong claws which may have been used for digging out roots and tubers.

even-toed ungulates is the Irish elk. Not only is its appearance truly magnificent – it stood 3m (10ft) high and the male had antlers with a rack spread of up to 3.3m (11ft) – but the number of fossils found in the Irish peat bogs indicate that it must have lived in impressively large herds. It was really a large deer, and like most deer it had antlers, not permanent horns. These are bony outgrowths from the skull which are used by the males for fighting in the rutting season, and then shed. New ones are grown for the following year. The Irish elk must have had to eat enormous amounts of food to enable it to grow such huge antlers every year.

It is difficult to know where to begin to describe the bovoids, the last and most numerous of our even-toed ungulate groups. Here are the antelope, cattle, sheep, goats and others, which together form such a feature of the modern world. And yet, their evolutionary history is quite short and can be explained quite simply. They have spread through the world since the Miocene, following the spread of the grasslands. They and their extinct forebears were well adapted to a savannah existence, with long legs to run and escape the predators and with complex teeth that could grind the grasses without wearing out. One of their most spectacular features has been the development of horns on their heads. In contrast to the antlers of the

deer, these were permanent projections of bone with a hard outer casing of horn.

The sub-ungulates – the elephants, sea-cows and their relatives – were an impressive group, even at the beginning of their evolutionary history. They appeared during the late Eocene, already recognizably distinct from the other placentals; so their origins are rather obscure.

Moerithium was one of the early forms. Found in Egypt, it was about the size of a pig, with thick legs that ended in broad feet with hooves on the toes. It had no trunk, no tusks, and did not look particularly like an elephant. However, its skeleton is of interest to palaeontologists because it shows certain features which are elephant-like, and others which are sea-cow-like, thus linking two groups of mammals whose relationship might not otherwise have been suspected.

From these early Egyptian sub-ungulates were derived two major elephant-like lineages. The first resulted in a great diversity of mastodonts and elephants; this group was highly successful throughout the Tertiary, and its numerous members soon spread through most of the continents. The second evolutionary line, the dinotheres, was quite different. From the Miocene to the Pleistocene its members remained virtually unchanged, except that later forms tended to grow larger, until they stood 3m (10ft) high. They were elephant-like with long trunks and peculiar down-curved tusks which grew from the lower jaws (elephant tusks are found in the upper jaws). The function of dinothere tusks has long been a subject for speculation; one curious early idea was that dinotheres were river-dwellers and that they used their tusks as anchors to stop themselves from drifting downriver in their sleep. Another more prosaic and more likely suggestion is that they were used for uprooting plants. Since these mammals are extinct, no one will ever be able to prove for certain the function of the tusks.

Alongside these herbivores evolved the placental carnivores, efficient hunters and killers. The bodies of these animals did not share the tendency to great size that is seen so often in the herbivores, and consequently they retained their mobility to a much greater degree. They had claws on their feet and these, together with the teeth, were used for attacking their prey. In many forms the canines, or eye teeth, were developed as killing fangs, and the carnassial side teeth were used as shears to slice the flesh from the carcasses.

The early carnivores, the creodonts, must have been rather slow, only marginally faster than the early ungulates which they pursued. Some were dog-like, others resembled cats or weasels, but most became extinct by the end of the Eocene, when the slow animals which formed their food were superseded by the later, faster ungulates. The creodonts could not catch these fleet-footed forms, and they could not compete with the modern carnivores either, which appeared on the scene at the same time.

These modern Tertiary carnivores have been characterized by a similarity of form which makes description of the fossil forms read like a textbook of biology. For instance, the cats evolved into the form in which they are found today at a surprisingly early date, and since the Oligocene they have changed but little. The dog lineages have varied rather more, and their diversification includes badgers, raccoons, wolves and others; all evolved from the early Oligocene forms. Sometime during the Miocene some of the dog-types became heavier in build and rather more catholic in their diet. Today their descendents are the bears, which figure so prominently in our colder climates.

For colder climates were on their way. The mammals of the Tertiary lived and diversified in a warm world for many millions of years, unaware of the gradual cooling that was occurring with the passing of the ages. The threat came to a head with the arrival of the Pleistocene ice age, when the world changed and the cold hostile regimes led, at the least, to a radical redistribution of both plants and animals and, at worst, to complete extinction of local populations.

Arsinotherium was a strange animal that lived in early Oligocene times. It was probably related to the sub-ungulates but nothing is known of its antecedents nor its descendents and it has several unique features which separate it from all other known mammals.

The Ice Ages

There are few climatic regimes as severe or as influential upon animal or plant life as ice ages. Ice-bound oceans and ice capped continents are singularly inhospitable; yet, for those few organisms able to survive along the margin of those refrigerated areas, they may offer large unoccupied ecological niches. The precise origins of ice ages are unknown; some scientists believe that fluctuations in the sun's energy output may be responsible, while others think that the earth shows some marked wobble upon its axis, which causes changes in weather patterns and ocean currents. Whatever the initial cause, it is clear that the formation of ice caps may affect areas very far removed from the poles. For instance, during the Pleistocene ice age so much water was bound up in the polar ice-sheets that the sea level dropped by 100m (300ft) thus exposing large areas of continental shelf. The marine animals and plants survived on the small remaining areas of shelf, but whole populations died out. Another characteristic feature was the marked differential between polar and equatorial temperatures causing a great increase in the number of storms and changing the pattern of rainfall in the tropics, with the result that some areas received no rain at all. At this time, the Sahara desert was formed.

Ice ages appear to be episodic phenomena, occurring approximately every 300 million years. There were four during the Precambrian, another during the Carboniferous and the last within the Pleistocene, some 2 million years ago. The Carboniferous glaciation is thought to have been in the southern hemisphere and, as in the Pleistocene, the temperature seems to have fluctuated. When it dropped, the ice caps spread and the level of the seas was lowered; at other times, the temperature increased, the ice caps retreated and the level of the seas was raised. Each cycle took several million years and the sequence of events was repeated a number of times; this phenomenon may be the explanation for what has long been a puzzle of the Carboniferous era. In the northern hemisphere at that time, the extensive coastal areas of swampland containing giant horsetails, huge clubmosses and large ferns were subjected to periods of inundation when the sea levels rose and the continental shelves became flooded. It can now be assumed that this flooding was due to a rise in the sea level, during an interglacial period in the ice age of the southern hemisphere.

It is the Pleistocene ice age which is of direct concern to us, since our present world is the immediate legacy of events which then took place. There is a common misconception that the onset of the ice regimes was instantaneous, but in reality it was a long, slow process. During the whole of the Tertiary, some 63 million years, temperatures were gradually falling; for example, the average European temperature in the Eocene was 21°C (70°F), in the Pliocene it was 14°C (57°F) and today it has fallen to 9 or 10°C (48 or 50°F).

It is possible that this fall in temperature was part of the reason for the spreading of the grassland in the middle Tertiary and this, in turn, greatly influenced the evolution of the mammals, as we have already seen. The Pliocene saw the peak of the mammal radiation – it was the last, great, warm epoch before the onset of the cold.

At this time, large areas of southern Europe were savannah country (grassland interspersed with clumps of small trees), while central and northern Europe was clothed with mixed forests of deciduous and coniferous trees, so typical of present day warm temperate regions. The plains were inhabited by huge ungulate herds, while the giant cats and the hunting dogs followed these herds; mammals are ideally adapted to savannah life and this was their heyday. Forests do not support such huge numbers of mammals, but nevertheless the woods were full of wild deer, pigs, tapirs and bears. In the rivers were hippos, beavers and others.

Common inhabitants of the plains were the mastodonts – descendents of *Moeritherium* and its relatives.

This Pleistocene mastodont has become trapped in a tar pool. These tar pools, which can still be seen at Rancho La Brea, California, acted as a trap for many different kinds of animals and have provided a perfect medium for the preservation of their bones.

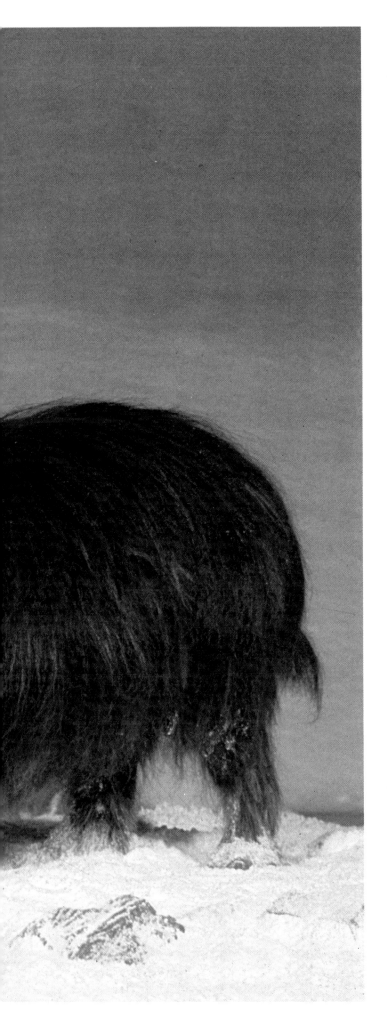

They were large relatives of the elephants, with long trunks and elongated upper and lower jaws, both of which had tusks; in the so-called shovel-tusked mastodonts of the Pliocene the lower tusks were very broad and formed huge shovels which could be used for uprooting plants. The first elephants appeared in the Pliocene, forerunners of the huge woolly mammoths of the ice ages.

The vast herds of ungulates attracted the great carnivores, among them the dirk-tooth cats – heralds of the sabre-tooths, and the hyenas. Modern hyenas are often considered cowardly scavengers, but the Pliocene short-faced hyenas and their relatives, the hunting hyenas were probably anything but cowards. The former were fearsome beasts, large as lions and able to pull down the heaviest of the ungulates; the latter were built like cheetahs, slight and fast, for running down the fleetest of the antelope.

Typical of the woodlands were the wild pigs. They probably led lives very similar to those of modern wild pigs but certain features of their feet suggest that they inhabited wet ground; if so, they would have seen the hippopotami relaxing in the slow-moving rivers. These large, semi-aquatic mammals are relatives of the pigs, and surprisingly, Pliocene European hippos were the same species as the modern African variety. At that time they were spread throughout Europe and Africa, but the ice age wiped out the European populations and only the African hippos live on unchanged.

As the temperatures dropped at the beginning of the Pleistocene, the climatic zones changed. The savannah lands gradually spread south until they were found only in Africa and southern Asia; the woodland spread south to take the place of the savannah, and arctic tundra or ice caps developed where the woods had been. The animals either followed the change in vegetation and migrated south, or died. However the Pleistocene was not continually cold – it was of the same periodic nature as the Carboniferous ice age. There were four cold glacial spells interspersed with three warmer interglacial times. It is quite possible that the ice age did not end 10,000 years ago, but that we are now living in another interglacial. As a glacial spell came to an end, the vegetational changes were reversed, the ice caps retreated and the northern latitudes were once more covered with forest or savannah.

Mammoths were enormously successful animals of the glacial times. Early forms were already present during the first glaciation, and the lines of evolution culminated in the great woolly mammoths, of the fourth interglacial spell. Unlike the mastodonts, the

The woolly rhinoceros of Pleistocene Europe and Asia had a long shaggy coat which effectively protected this animal against the cold climate of that time.

This reconstruction of a Pleistocene scene during a glacial spell, shows woolly mammoths moving across a snowy landscape. These animals were a major source of food for Neanderthal men.

mammoths were true elephants, having shortened jaws and only one set of enormous curved tusks in the upper jaws. The abrasions on the lower surface of these tusks suggest that the animals probably used them for scraping away the snow to reach the grass beneath. The woolly mammoths spent the summers on the tundra, migrating into the forest belt in the autumn. Their size must have ensured that they had few enemies, but clearly early man was a bitter foe. Cave drawings depicting mammoths are numerous and the animals were slaughtered in great numbers for food.

Several carcasses of these mammals have been found frozen in Siberia; the most famous was discovered in 1900 and is now exhibited in Leningrad. It had fallen into a fissure, where it must have been killed instantly for it still had grass between its teeth. Since the bottom of the crevasse was within the permafrost layer, the animal was gradually frozen. As can be imagined, the finding of a whole mammoth has given us far more information about the life of these impressive beasts, than any skeleton ever could have done. If we did not know that the mammoths were covered in shaggy skin,

The most common carnivore at the Rancho La Brea tar pools was the Dire Wolf—to date 1646 individuals have been excavated. This great carnivore was as big as a large modern wolf but had a larger head and was heavier with shorter legs. With this build and a small brain, it may have been more suited to a scavenging rather than a hunting role, similar to that of present day hyaenas.

we might be surprised to find an elephant so far north, in an ice age. Equally surprising would be the discovery of Siberian rhinoceroses from the third glacial spell, if the frozen carcasses of these animals had not been preserved intact with their woolly skins. Rhinos used their horns to scrape away snow, just as the mammoths used their tusks.

These large animals, together with musk oxen, reindeer and polar bears, lived in the shadow of the ice caps. During the glacial times they could roam the whole of northern Europe, but in the warmer interglacial spells, they were confined to the northernmost fringes of the land mass, while the more southerly areas were once more invaded by the inhabitants of the temperate climes. Straight-tusked elephants,

ARCTIC CIRCLE

Map showing maximum extent of the glaciation in the northern hemisphere during the Pleistocene epoch, as seen from the North Pole. The ice caps cover not only Greenland and Iceland but also large parts of Europe, Asia and North America.

hippopotami, horses, deer and cattle moved north as the snow line retreated. Following the herds were the great carnivores, the hyenas, cave lions and the scimitar cats.

These scimitar cats were peculiar for several reasons. Their front legs were long, but their back legs were quite short and the whole of the foot was in contact with the ground. In modern cats only the toes touch the ground and this adaptation gives them the fleetness of foot necessary for catching their fast-running prey. (Try running fast flat-footed and you will see how difficult it is.) The flat-footed stance of the scimitar cat in conjunction with its size – it was as large as a lion – must have made it very slow-moving. The teeth of the cat were very sharp, but quite thin and delicate,

and this feature, together with its slowness, probably made its choice of prey quite limited. The scimitar cat became extinct by the end of the Pleistocene.

Another Pleistocene carnivore was the European cave bear, in many ways the climax of bear evolution. They were massively built animals, with stub-like tails, great raking claws and heavy teeth obviously better adapted to an omnivorous diet than to a carnivorous one. As suggested by their name, these bears lived in caves, where their bones have been found in their thousands in the cave deposits of central Europe.

All sorts of animals expect to find shelter in a cave, like the cave bears, but the richness of fossil cave deposits, from the ice age, testifies to the fact that they often found death instead. Some animals have been found obviously buried by a fall of rock from the roof – death and fossilization in the same instant! The large numbers of diseased skeletons, from all kinds of mammals, suggest that caves provided a last resting place for dying animals – places where they could rest without fear of harassment from predators. Cave deposits are often particularly rich in fossilized infant

bear bones, suggesting that they died in hibernation, presumably because the mother died.

In North America, a most unusual form of fossilization has given us a great deal of information about life on the southern part of that continent. The Pleistocene tar pools of California were situated in a temperate zone and they provide us with a unique cameo of life, away from the ice. These pools were death traps to the animals that ventured into them and the tar formed a perfect medium for the preservation of their bones. Often a thin film of water covered the surface of the pools and animals lured into drinking became inextricably bogged down.

The herbivores include not only the usual assortment of horses, mammoths, pronghorns and bison, but also examples of the South American ground sloths and glyptodonts. Obviously, by this time the connection between South and North America had become established and migration was occurring. The North American bison were most impressive, standing over 2m (6ft) high and having a much wider horn spread than existent forms.

However, it is for their unique collection of carnivores that the tar pools are justly famous. In a normal population, the ratio of herbivores to carnivores is a high one, many herbivores being needed to maintain one carnivore. In the tar pool deposits, the total number of carnivores actually exceeds the number of herbivores, presumably because large numbers of wolves, coyotes and cats were attracted by the trapped and dying animals. The most impressive of these must have been *Smilodon*, the greatest of the sabre-toothed cats. *Smilodon* was as large as a lion, relatively slow-moving, and attacked large, slow-moving prey such as the mastodonts, killing them with huge, dagger-like canines. Modern cats use both upper and lower canines to grip and kill their prey, but *Smilodon* stabbed its prey, probably in the back of the neck. Scavenging and predatory birds are also unusually common in the tar pools, the tar providing one of the few media soft enough to preserve the fragile bones without crushing them.

The last great mystery of the Pleistocene was, as is so often the case, one of extinction. Many of the animals described survived the fluctuations of temperature of the first three glacial spells and the three interglacial times, and then became extinct in the fourth glacial. Why was this last glacial spell different from the others? It is probable that at least part of the answer to that question lies in the following chapter.

This sabre-tooth cat was one of the great carnivores which hunted the vast herds of ungulates living on the plains of the Pliocene epoch. It had enormous canines with which it stabbed its prey, probably in the back of the neck.

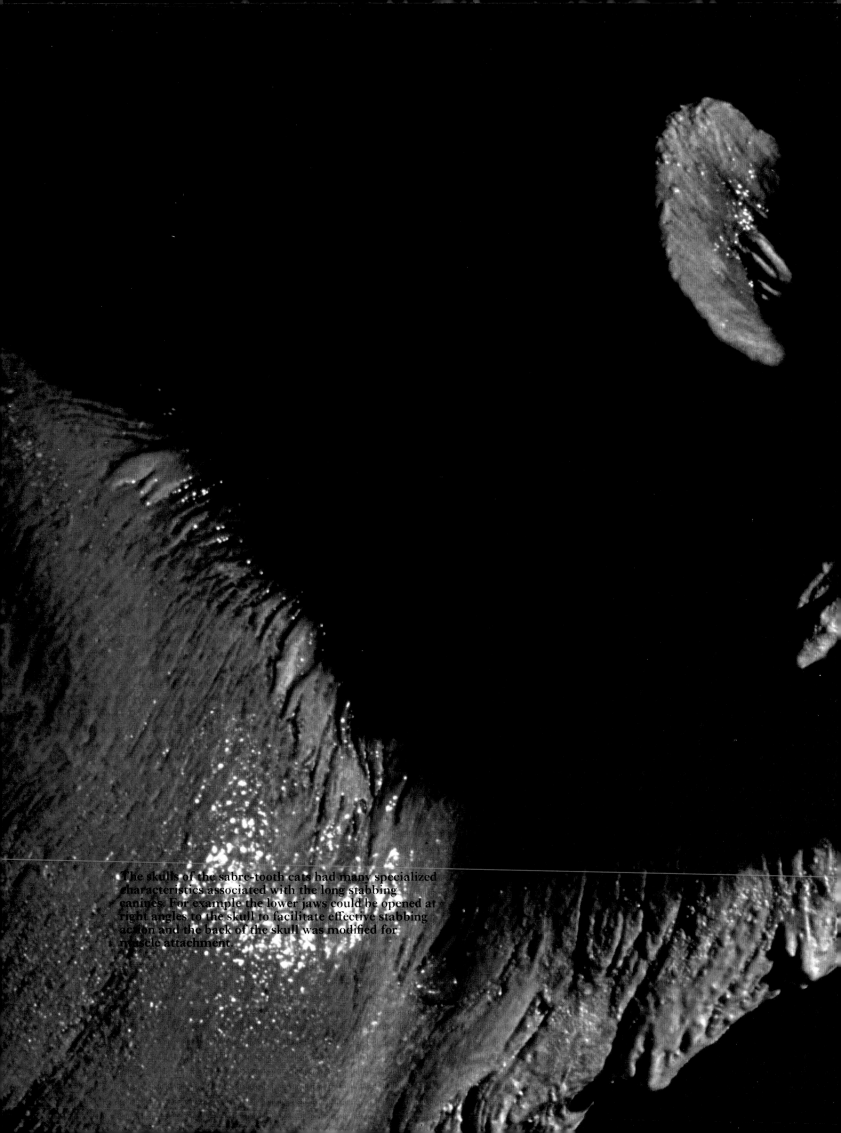

The skulls of the sabre-tooth cats had many specialized characteristics associated with the long stabbing canines. For example the lower jaws could be opened at right angles to the skull to facilitate effective stabbing action and the back of the skull was modified for muscle attachment.

The Emergence of Man

During the 19th century two books by Charles Darwin revolutionized our thinking about the orderliness of the natural world, and especially about man's place in nature. *The Origin of Species*, published in 1859, proposed that the present day diversity of life can be explained by the descent of animals from other animals, and that Special Creation was not a viable theory. *The Descent of Man*, published in 1871, treated mankind purely as another species of animal, subject to the same universal laws of life, death and change. This created a great sense of outrage amongst the Victorians, steeped in contemporary Christian dogma which recognized man as above animals and not subject to their rules of life. But, more particularly, many Victorians took exception to Darwin's view that man and the great apes were descended from a common ancestor. One must remember that the literate in Victorian society were prepared to pay handsomely to ridicule chimps at the zoo. Darwin was proposing that these animals were our nearest living relatives! Today the idea that *Homo sapiens* or modern man is one of the primates has gained general acceptance.

The primates diverged from the basic insectivore stock very early in mammalian history, perhaps even during the Cretaceous period. The primitive forms were small, about squirrel size, and led arboreal lives similar to those of their insectivore ancestors. They had several distinctive features, amongst which were manipulative hands and stereoscopic vision – two conditions which, when present together in one animal, led to great dexterity in handling objects. The present day Oriental tree shrews are similar to the early forms, and they exhibit another primate characteristic – that of an extended learning period in the juveniles, when the association between parents and young becomes cemented into a true family life.

From the primitive stock evolved the lemurs, a group that was most successful in the Eocene, and

Hominids belonging to the species *Homo erectus* lived in caves which they kept warm by lighting fires. They made extensive use of tools including hand axes, choppers and flakes; these latter tools were used for skinning animals such as the deer shown.

which survives today on the island of Madagascar. It is thought that the two major monkey lineages evolved from some of the lemurs, either separately or from one common ancestor. The New World monkeys, found today in South America, have flattened noses and prehensile or grasping tails, while the Old World monkeys from Asia and Africa have protruding noses and non-prehensile tails. It is from the ancestors of this latter group that the great apes and man are thought to have been derived.

The ape lineages include the gibbons, orang-utans, gorillas and chimpanzees, our closest living relatives. Features that distinguish them from all other primates include long arms that were used primitively for swinging through the trees, and the lack of a tail. The fossil record of apes is as fragmentary as that of early men, but one particularly famous form was *Proconsul* (now included in the genus *Dryopithecus*). It was named after a chimpanzee, Consul, which lived at the London Zoo at the time when the remains were discovered, in 1926. *Proconsul* lived about 18 million years ago, in the early to middle Miocene, just before the apes and man diverged.

There is a certain amount of controversy as to the order in which the human features appeared, but the latest views are that one of the first was the change from a crouching ape-like posture to an upright man-like position. This involved changes in all parts of the skeleton, from the skull where the opening for the spinal column moved to the base from the back of the skull, to the feet, which formed a stable platform to bear the whole weight of the body in man. This is in contrast to the original hand-like grasping structure of the feet of the apes.

Assumption of the upright posture indirectly influenced the evolution of the hands and brain. Since the former structures were no longer used in locomotion, they could then be used in other ways; the evolution of the opposable thumb led to the power grip in early men, enabling them to make tools and club-like weapons. Later men developed a precision grip, in which the fingers and thumb were proportioned to facilitate exact gripping of precision tools or

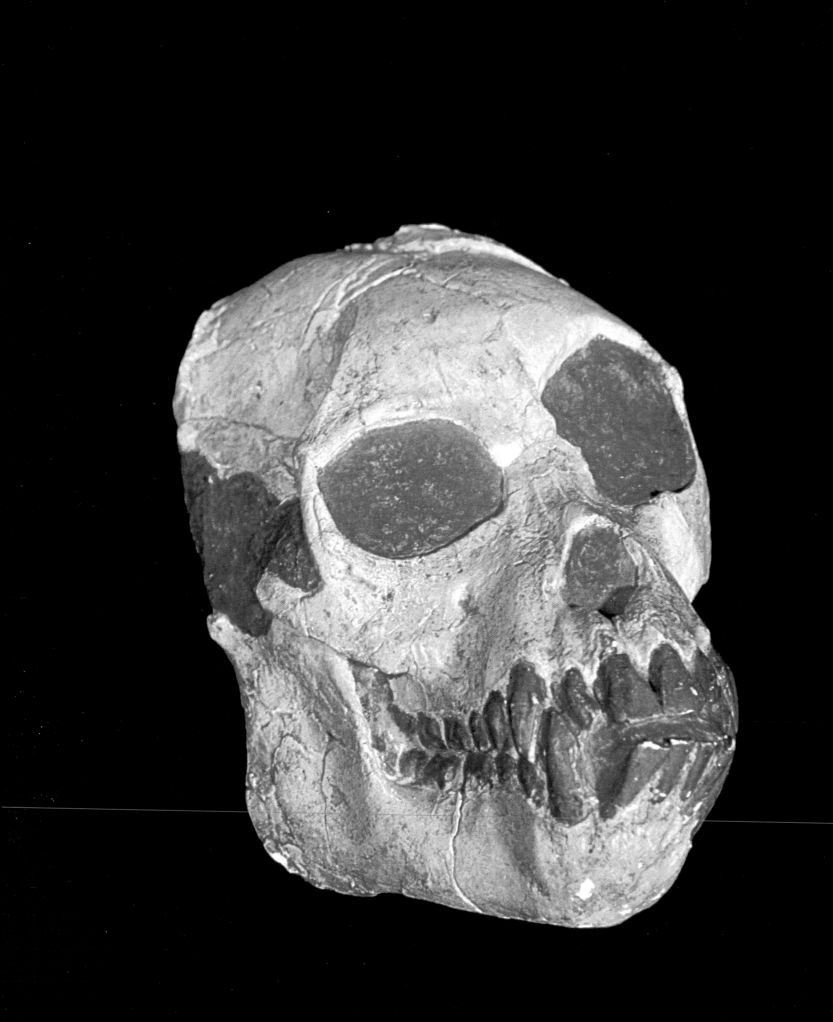

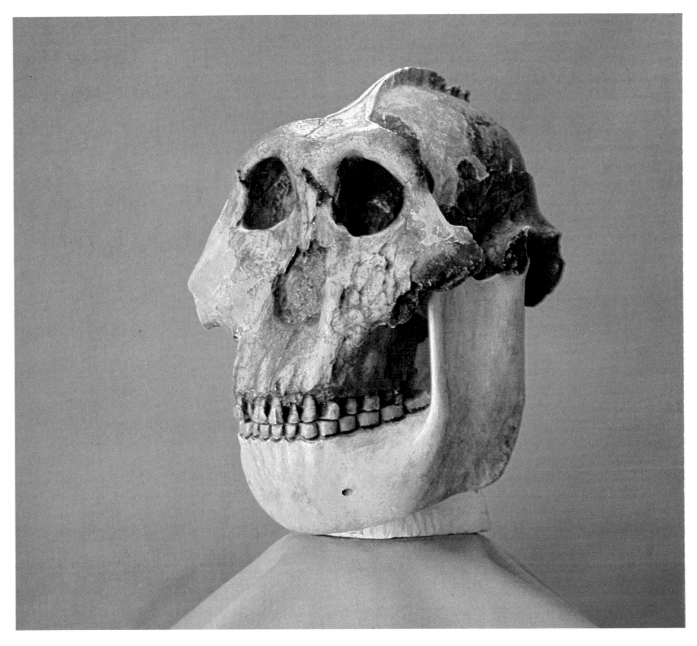

Left: This skull came from an ape called *Proconsul* which lived in East Africa during Miocene times. The skull had a low rounded braincase, distinct eyebrow ridges and a projecting face and lower jaw.
Above: The original *Zinjanthropus* skull was discovered in Olduvai Gorge in 1959. It is now known to be an australopithecine whose distinguishing features include a deep face with massive molar teeth.

weapons. This evolution of manipulative ability is thought to have occurred in conjunction with the enlargement of the brain, another trend seen in the evolution of man. The increase in brain size probably happened through a process of neoteny, a situation in which sexual maturity was delayed beyond the normal time. This led to a retention of infant characteristics including the large juvenile brain size. Of course, these human features took many millions of years to evolve, and it was only about 40,000 years ago that Cromagnon man, the first of the modern men, appeared.

The very early hominids such as *Ramapithecus* looked much more like apes than men, and indeed, some experts think that they were apes, not men at all. Jaw fragments from this animal have been found in Africa, Europe and India, and it was the remains from the latter continent that gained it its name – the ape of the god Rama. Its relationship with man is based on jaw shape and teeth alone. Its canines are more like those of men than ape canines. Until more is known about the rest of the skeleton, its evolutionary position, with respect to the origin of man, remains a mystery.

Ramapithecus lived during the Miocene between 10 and 14 million years ago; unfortunately no hominid remains have been found between that time and the Pliocene, when *Australopithecus* appeared, about 4 million years ago. In 1924, a young professor from Johannesburg, Raymond Dart, received the cast of a skull plus several bone fragments from a lime mine at Taung, near Kimberley, South Africa. On examination, they proved to belong to an infant hominid which Dart named *Australopithecus africanus*, the Southern

ape; he suggested that it belonged to a new group of animals which he called ape-men. Considerable scepticism greeted this idea, but Dr. Robert Broom, a retired medical doctor attached to the Transvaal Museum at Pretoria, was receptive to the proposal, and he set out to find more evidence. He was successful in 1935, when he discovered skulls and teeth in a cave at Sterkfontein, in South Africa. This cave has since proved to be a rich source of material for *Australopithecus africanus*, the so-called gracile australopithecines, and a description of these ape-men is now possible. They were about 1.5m (4.5ft) high, with an upright posture and a brain capacity of 480cc. When considered in relation to their size, this brain capacity was about halfway between that of a chimpanzee (410cc) and that of a present day man (1,400cc). However, little is known of their way of life and no tools have been found with their remains.

Another australopithecine was found by Dr. Broom in 1938. From a cave at Kromdraai, near Sterkfontein, came skulls and teeth; later finds from other South African localities have produced limb and girdle bones. These remains were given the name *Australopithecus robustus*, or robust australopithecines. They differed from the Sterkfontein fossils in that they had larger teeth and stouter skulls, features that have given

Above: **The primitive hand axe on the left and the two pebble choppers are 2 million years old.**
Right: **This skull, known as 1470 was found at Lake Turkana in 1972. It's owner lived 1.25 million years earlier than *Homo habilis* but had a larger braincase. Obviously there were several parallel lines of hominid evolution in which brain size increased at different rates; only one led to modern man.**

them their name. Like the gracile forms, they were upright in posture, but they were taller and had larger brains (500cc).

The relationship of these australopithecines to each other and to modern man is still in doubt. One possibility is that the older species, *Australopithecus africanus*, gave rise to *Australopithecus robustus* and *Homo* (the genus to which modern man belongs) as separate, later lineages. This theory is supported by evidence from Swartkrans, South Africa, where in 1949 Dr. John Robinson found *Australopithecus robustus* remains alongside skull fragments that appeared to belong to *Homo*. If the two species were alive at the same time, then their relationship could be that of cousins, but their derivation from *Australopithecus africanus* remains unproven.

The story was further complicated in 1959, when a new species of *Australopithecus* was discovered. This was a skull found in Olduvai gorge, East Africa, by

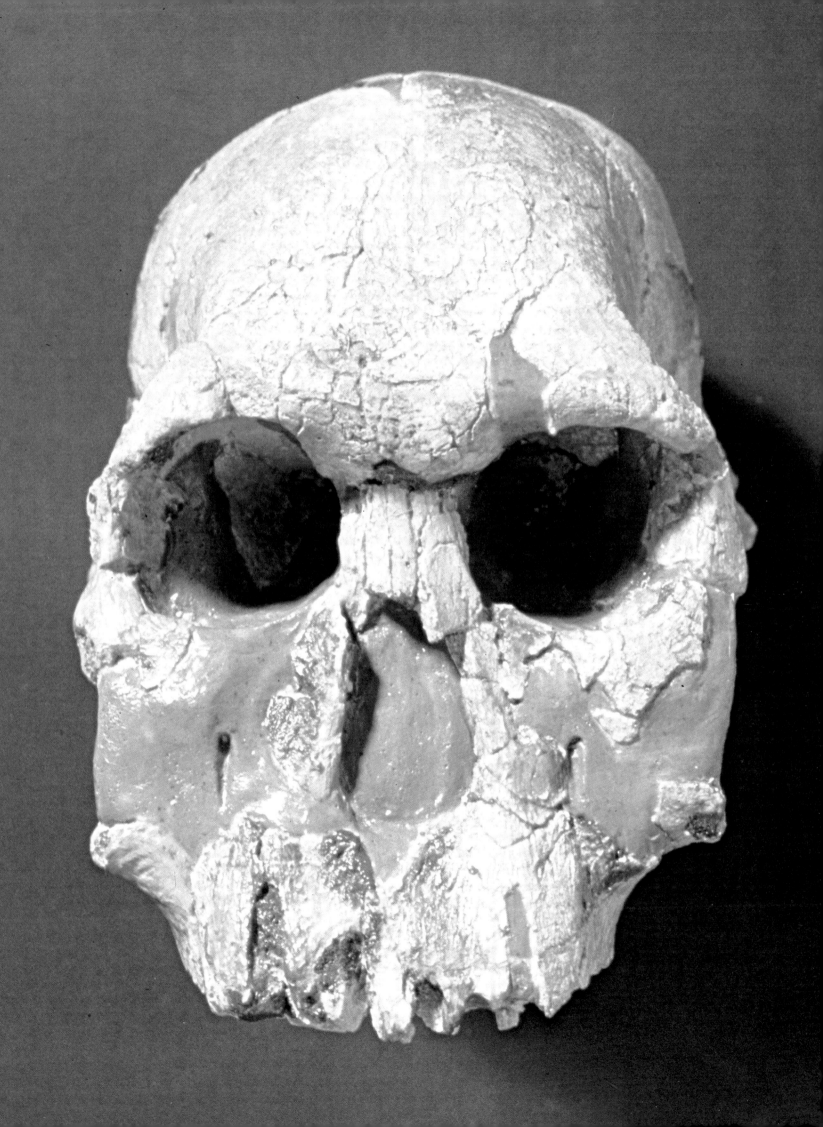

Mary Leakey, wife of Louis Leakey. This husband and wife team had been searching the area for about 25 years, and experts were sceptical about the Leakeys' belief that hominid remains would eventually be found. The skull, then, represented a great victory for the Leakeys. Initially it was named *Zinjanthropus* or 'nutcracker man', but later, experts realized that it was very similar to *Australopithecus robustus*, and it was given the name *Australopithecus boisei*.

This find led to extensive new excavations in the gorge, but instead of further australopithecine remains, the search unearthed another more advanced hominid which was given the name *Homo habilis*. Early members of this species probably lived at the same time as *Australopithecus boisei*, about 2–2.5 million years ago; but in contrast to the more primitive form, they made simple tools by chopping a single sharp edge into a rounded stone. Later members of *Homo habilis* made more effective artefacts by chopping two flakes from the original stone, thus creating much sharper edges to their tools. They probably also made crude shelters, for at Olduvai, Mary Leakey found the remains of a stone circle, about 1.75 million years old. The stones could have been used to anchor skins or other covering materials over a framework of sticks.

The remains of *Homo habilis* included hand bones, and these were sufficiently different from the ape-like condition to suggest that a power grip was used, confirming the view that the tools and shelters were made by this particular hominid. Foot bones, present at the site, showed that the feet had evolved into stable platforms, similar to those of modern man and able to sustain the weight of the creature, although the arches were smaller than in present day man. The skull and teeth were more like those of modern man than those of the australopithecines, and the brain size (about 680cc) was definitely larger than that of any of the earlier hominids. Some people have suggested that *Australopithecus* lived in troops, much like modern baboons, whereas *Homo habilis* lived in family groups, building their crude shelters and making their tools. Studies made by Richard Leakey (a son of Louis and Mary Leakey) at Lake Turkana in Kenya seem to confirm that the two hominids lived in these different ways.

An extension of this family way of life was characteristic of men living between 1 million and 0.25 million years ago. This new species, *Homo erectus*, lived not only in Africa, but also in Indonesia, Europe and Asia. The earliest finds were made in Indonesia, where the so-called Java man was discovered in 1891. The relationship of this missing link between man and

The Neanderthal face was rather different to that of other men. There were prominent brow ridges and swept back cheek bones accentuating the large, oddly protruding braincase.

Neanderthal families lived in European caves during
the fourth glacial spell. They were efficient hunters and,
using the skins as clothing together with fire, they
survived the long cold winters.

ape remained obscure, until later finds in Indonesia and China (Peking man) revealed more of its anatomy. The excavations at Peking, which began in 1921, were particularly interesting; a series of caves in Dragon Bone Hill had obviously been the sites of hominid homes for many generations. Both cutting and chopping tools were found, together with the remains of deer and other animal bones – obviously the remnants of ancient meals. An entirely new feature was the use of fire; many old hearths were present in the caves.

Right: This Cro-Magnon woman was buried at St Germain, France between 10–20,000 years ago. She was found in the crouching position, under a small stone which is shown behind her, together with flint implements, a dagger of antler and a necklace of animal teeth.

Below: Restoration of the head of a Cro-Magnon man.

Cro-magnon men had replaced Neanderthal men on the Russian steppes 25,000 years ago. These modern men hunted with spears and traps and their shelters of mammoth tusks, bones and skins indicate that these animals were prime targets for these hunters.

Other members of this species have since been found in Europe (the famous Heidelburg jaw belongs to *Homo erectus*), and one cave site in southern France shows evidence of the building of shelters with stone circles, like those described for *Homo habilis*. These French people were obviously living a cooperative life, for the camp sites were large, with many stone tools present; large numbers of animal bones suggest that they hunted in a group, especially since some of the prey were too large for one man to kill alone. *Homo erectus* has also been found in Africa, at Olduvai, Lake Turkana and the Southern African sites, in later beds than the australopithecines or *Homo habilis*. At Olduvai, large stone hand axes appeared at the same time as the earliest *Homo erectus* fossils, and it seems likely that it was these men that invented them. The hand axes were large cutting tools, and they are found in the rocks with a great variety of other tools such as choppers and cleavers – far more than those associated with *Homo habilis*. The axes are also found with *Homo erectus* remains in Europe.

What did *Homo erectus* look like? They were man-like in appearance with a long skull and medium-sized brain (about 950cc) but no forehead. They stood about 1.5m (4.5ft) high, and they probably walked with rather a stooping gait. Most experts think that they were the forerunners of *Homo sapiens* because, about 0.25 million years ago, men intermediate in form between *Homo erectus* and *Homo sapiens* appeared, of which perhaps the most famous is Swanscombe man, the remains of which were found near London in 1935.

One of the most controversial subjects with regard to the beginnings of *Homo sapiens* is the relationship of modern man with *Homo neanderthalis*. These often caricatured men lived in Europe throughout the fourth glacial period of the Pleistocene. They were characterized by odd bun-like protuberances at the backs of their skulls, swept back cheek bones, massive eyebrow ridges and a very robust build. They were shorter than modern men. Unfortunately one of the first specimens described was arthritic. This led to the erroneous idea that they were shambling in gait, and their peculiar faces have given them the reputation for being stupid. However, their brain cases were as large, if not larger (average 1500cc), than those of modern men, and they manufactured a large variety of tools, including scrapers, knives and advanced hand axes. These are not the characteristics of a stupid race. In addition, they were skilled hunters who worked hides for protection from the bitter ice age winters. Not only were the skins used for clothing, but also for shelters erected on sticks within their caves. The neanderthalers may have been partly responsible for the extinction of many of the large ice age mammals that died out during the fourth glacial time.

Neanderthaloid-type remains have also been found in the Middle East and North Africa, but these members of the species show differences from the European stock, notably in the absence of the bun-like protuberance. The whole question of the relationship of all these men to *Homo sapiens* remains unclear. They may have been members of another species, derived separately from *Homo erectus*; alternatively, they may have been just another racial variant of *Homo sapiens*. At the end of the Pleistocene, neanderthal men became extinct. If they were a variant of *Homo sapiens*, then presumably they could interbreed with the more advanced Cro-magnon men, who were spreading across Europe at that time, and thus became absorbed into later populations. The other more sinister explanation is that the neanderthalers were wiped out by the Cro-magnon migrations.

The Cro-magnon men gain their name from the place where they were first found in France, in 1868. They appeared about 40,000 years ago, originating in Africa and Eurasia, later invading America and Australasia. They were modern in all respects, with large brains (1400cc), protruding chins and the stature of modern men. Their culture involved the use of fire, advanced stone tools such as spears and axes with wooden hafts, and complex hunting techniques using pitfalls and traps. These men probably completed the extermination of the larger ice age mammals, such as the mammoths, a process which had been started by the neanderthalers. They lived in wooden huts or caves and, like the neanderthalers before them, they buried their dead with ritualistic ceremonies. This burial pattern implied a degree of awareness of the mystery of life and death, a premise further reinforced by the development of artistic ability. Cro-magnon cave paintings were probably symbolic rather than decorative, for they were found in the deepest recesses of the caves. Sculpting also began at this time, and small carved figurines of the female form (almost certainly fertility symbols) have been found at localities all over Eurasia.

These were the first of the modern men, the real *Homo sapiens*. We now pass from prehistory into history, into the development of agriculture, which in turn was responsible for the slow change from a nomadic hunting life into one with more stable, settled communities. The total populations gradually increased, and as the climate improved with the retreat of the ice, then the men became an ever more significant factor in the ecology of this planet.

Above: **Primitive carvings included not only female figurines but also animal sculptures. This figure of a bison comes from Dordogne, France and was carved from an antler about 12,000 B.C.**

Below: **Cro-magnon men are famous for their cave paintings, many of which are found in France and Spain. This one, from a cave at Rouffignac, Dordogne represents a mammoth with ibexes.**

Index

Page numbers in italic refer to illustrations.

ACKNOWLEDGEMENTS

The publishers would like to thank the following
individuals and organisations for their kind permission
to reproduce the photographs in this book:–

The American Museum of Natural History 50, 98 left;
Ardea London (P. J. Green) 11 above and below, (Arthur
Hayward) 14, 22, 24, 29 above, 37, (P. J. Green) 42,
(Arthur Hayward) 43, 47, (P. J. Green) 52 above left, (Ian
Beames) 54, (P. J. Green) 58, 60, (Arthur Hayward) 64,
(P. J. Green) 74, (P. Morris) 76, (Arthur Hayward) 78,
79; Bruce Coleman Ltd (Eric Crichton) 6, (R. I. M.
Campbell) 91; Deaton Museum Studio (St Paul Science
Museum, Minnesota) 66–67; Mary Evans Picture
Library 7; Imitor Ltd 17, 52 below, 69, 75, 82, (British
Museum (Natural History) 15, 23, 52–53 above right, 57,
68, 72–73, 90; Institute of Geological Sciences 16 right;
Jacana (Alain Cropt) contents, (Jean-Philippe Varin)
28–29 below, (Alain Cropt) 61, 65, (Jean-Philippe Varin)
86–87, 94–95; Natural Science Photos (Arthur Hayward)
end papers, title page, (F. Greenaway) 25, (G. Kinns) 39,
(Arthur Hayward) 46, (Arthur Hughes) 48–49, (Arthur
Hayward) 62, 70 below, (G. Kinns) 84–85; N. H. P. A.
(H. R. Allen) 38; Photoresources 20, 21, 98 right 99, 103
above and below, (British Museum (Natural History)) 16
left, 26–27, 92 right; Rida Photo Library (B. Wood) 93;
Tate Gallery, London 8–9; John Waechter 92 left and
center; Zefa (Clive Sawyer) 40–41.

The publishers would also like to thank the following
artists for supplying the illustrations:

Andrew Farmer 12–13; David Nockels 18–19, 44–5,
80–1, 88, 96–7, 100–1; Ralph Stobart 24, 34, 55, 56, 63;
George Thompson 30, 31, 32, 59, 70 top; Brian Watson
(Linden Artists) 83.

PDO 79-181